T0136823

PERRY'S

Arcana

Arcana,

OR
MUSEUM OF NATURE;

CONTAINING

Delineations of the most recent Discoveries,

ON THE SUBJECT OF

NATURAL HISTORY:

WITH

CLASSICAL AND DESCRIPTIVE EXPLANATIONS.

LONDON:

PRINTED BY G. SMEETON, ST. MARTIN's LANE,

FOR

J. STRATFORD, 112, HOLBORN HILL.

[PRICE HALF-A-CROWN.]

PERRY'S

Arcana

A Facsimile Edition

With a Collation *and* Systematic Review

RICHARD E. PETIT

THE
ACADEMY
OF NATURAL
SCIENCES

TEMPLE UNIVERSITY PRESS
PHILADELPHIA

Temple University Press
1601 North Broad Street
Philadelphia, Pennsylvania 19122
www.temple.edu/tempress

Digital restoration by Paul Callomon

This work may be cited as Petit, R. E., 2009. Perry's Arcana: a facsimile
edition with a collation and systematic review. The Academy of
Natural Sciences and Temple University Press, Philadelphia.

This book is printed on acid-free paper for greater strength and longevity.

Library of Congress Cataloging-in-Publication Data
Perry, George, b. 1771.
  [Arcana, or, The museum of natural history]
  Perry's Arcana. -- A facsimile ed. / with a collation and systematic
review by Richard E. Petit.
      p. cm.
Originally published in monthy installments, under the title Arcana, or,
The museum of natural history. London : Printed by G. Smeeton for James
Stratford. 1810-1811.
  Includes bibliographical references and index.
  ISBN 978-1-4399-0195-3 (cloth : alk. paper)
1. Natural history illustration. 2. Animals--Pictorial works. 3.
Plants--Pictorial works. 4. Perry, George, b.1771. 5. Natural history
illustrators--Great Britain. I. Petit, Richard E. II. Title. III. Title:
Arcana.
  QH46.P48 1810a
  590.22'2--dc22
                                                              2009023265

Printed in China

2 4 6 8 9 7 5 3 1

CONTENTS

FOREWORD

———

By Paul Callomon

GEORGE PERRY was a gentleman student of natural history in Regency England who produced two of the finest illustrated works of his time. Apart from this, almost nothing is known about him. He was born in 1771, and spent his early life in Shropshire and Liverpool before moving to London in 1807. An architect and stonemason by profession, he also lectured and wrote about natural history with considerable energy and enterprise. Between 1810 and 1811 he published the *Arcana,* a magazine of natural history, and in 1811 the *Conchology,* a large illustrated book on shells. Immediately thereafter, however, he returned to obscurity and no record can be found even of his death.

In the late eighteenth and early nineteenth centuries, London was the center of the scientific world. From the frontiers of Britain's growing empire came a ceaseless flow of new animals and plants to delight and puzzle men of science, who formed clubs and societies before which to present their findings. Many of the leading figures of the age were "amateurs", a term that in those days bore no negative connotation, and Perry was in no way unusual. In the heart of the city, "menageries"—private zoos—were operated for the education and entertainment of the public, stocked with live and preserved specimens imported by dealers. Parts of these collections later found their way into the natural history museums of the Victorian era, and many of their specimens survive today.

The advent of this wealth of new creatures coincided with explosive growth in the printing trade, and the result was a great bloom in scientific publication. There were large, lavishly illustrated books for those who could afford them, but some works were instead published in installments. This magazine format allowed readers to spread their cost or to select only those parts that interested them and to bind them as they wished. Complete sets of such works naturally tend to be scarcer nowadays than is the case with single-volume books, and Perry's *Arcana* is no exception. His work was furthermore unjustly spurned by many naturalists of its day, for reasons at which one can only guess, and many sets probably were later plundered for their fine figures. An exhaustive search, detailed herein, has so far uncovered only thirteen complete sets worldwide.

The *Arcana* reflects the views and enthusiasms both of its author and his audience. Descriptions of new and fascinating animals in often florid prose are presented alongside travelers' reports from the lands of their origin. The eclectic content nevertheleless maintains a fine balance between scientfic examination and more general observation, and the level of conjecture is no higher than in many more serious and elaborate works of the time. In the spirit of its age, Biblical references and early evolutionary theory sit together in amiable co-existence. Above all, the figures delight the eye with their imaginative arrangements and assured draftsmanship, the majority having been sketched from actual specimens. Many of the new scientific names Perry introduced in the *Arcana* remain valid and all are listed here with their present-day treatments.

With the two hundredth anniversary of its publication fast approaching, the fortunate availability of a complete, disbound set of the *Arcana* has provided a wonderful opportunity to bring Perry's work back into the public eye. Digital scanning and cleaning have restored glowing life to the often dramatic figures, and Mr. Petit's rigorous and erudite analysis guarantee that this new facsimile will be of considerable importance to natural historians in the twenty-first century.

TECHNICAL NOTES

As Mr. Petit asserts, the typography in the *Arcana* is quite good. The general quality of text printing, however, was not. For this facsimile, the text and plates were digitally scanned and cleaned to remove foxing, stains and dirt. Pages that had been heavily skewed during printing were straightened, and "shadowing" caused by ink leaching between pages was removed. Naturally no attempt was made to correct the content, though errors in spelling and grammar abound.

The colors of the plates were restored on the assumption that they were originally printed on off-white lithograph stock. This would not have been as bright as modern paper, and care has thus been taken not to overbrighten the colors. The sole botany illustration in plate 2 seems to have a strong green cast, but this was apparently the artist's intent.

Finally, Perry re-issued two numbers of the *Arcana* in order to standardize the layout so that each new part would start on a recto page. As new scientific names were introduced in both numbers, however, it was decided to reproduce the original versions here.

PERRY'S
Arcana

COLLATION AND
SYSTEMATIC REVIEW

INTRODUCTION

Fʀᴏᴍ Jᴀɴᴜᴀʀʏ 1810 to September 1811, the British architect and amateur naturalist George Perry issued a lavishly illustrated work on natural history in monthly parts. It was titled *Arcana; or the Museum of Natural History: containing the most recent discovered objects. Embellished with coloured plates, and corresponding descriptions; with Extracts relating to Animals, and remarks of celebrated travellers; combining a general survey of nature.* The *Arcana* ran to 84 plates and more than 300 pages of text treating mammals, birds, reptiles, fish, mollusks, echinoderms, insects, trilobites and one tree. Although it contained the first published illustration of the koala and descriptions of many genera and species that Perry supposed to be new, the work was forgotten for a hundred years until called to the attention of the scientific community by Mathews & Iredale in 1912. The *Arcana* has since lapsed again toward oblivion and many of the names that Perry proposed have been ignored by modern workers.

Although the rarity of this work partially accounts for its obscurity, it is also likely that it was actively ignored or perhaps even suppressed by members of the scientific establishment in England. Perry's *Conchology*, published in 1811, met a similar fate. There is a documented historical record of later animosity to that work shown by well-known conchologists such as Sowerby and Reeve. That the *Arcana* was ignored, however, cannot be blamed entirely on workers on mollusks, as new taxa were proposed in several phyla. The *Conchology* was listed by Brown (1816) in his list of "the principal books which treat of Conchology", but none of Perry's taxa are mentioned in his work. Perhaps the non-acceptance of the *Arcana* was at least partly due to the influence of Sir James Edward Smith, founder of the Linnean Society of London, who was at the height of his power when Perry was publishing. Perry praised Lamarck in his writings, and in the very first issue of the *Arcana* proposed new genera of Mollusca and indicated that the number of Linnaean genera in botany should be increased. These actions would not have been viewed favorably by adherents to the Linnaean classification.

The facsimile edition of the *Arcana* offered herein is intended to draw attention to Perry's work and his classic illustrations. Only recently have some of Perry's unused names that are senior synonyms of current names formally been declared to be *nomina oblita* (forgotten names that can no longer be used), five molluscan taxa of Perry's *Conchology* being so treated by Petit (2003). It is hoped that by drawing attention to other such forgotten Perry names, specialists in various fields of systematics will be encouraged to determine their proper status, and thereby complete the integration of Perry's work into the history of discovery of the natural world.

DATE AND FORMAT

The *Arcana* was issued in monthly parts, each consisting of four colored plates and sixteen pages of letterpress (two signatures), plus covers. The plates are dated on the first of the month in which the issue appeared. Mathews and Iredale (1912) showed that the dates of parts that could be matched to other works were correct and there is no reason to believe that they appeared at any time other than the printed date. Exceptions are plate 66, which is imprinted May 1, 1810 but was issued May 1, 1811, and plates 62 and 64, both dated March 1, 1811 but issued April 1, 1811. The existing covers are dated, beginning with number IV, and confirm the stated dates.

Neither the plates nor the pages are numbered, with the exception of the plates in the first issue, each of which is engraved as "Pl. 1" with the subject at the top of the plate. Mathews and Iredale numbered the plates consecutively using Roman numerals. The usage of Roman numerals was continued by Petit (2003) in order to prevent confusion with the *Conchology* plates treated in the same work. However, Arabic plate numbers are used herein to conform to current usage. Due to the absence of page numbers it has become customary to follow Sherborn (1922–32) in using signature numbers for page designations. This is possible as the first page of each signature is denoted by a letter or letters. There are some exceptions to both the order of plates and their numbers as given by Mathews and Iredale and also to what would be assumed to be the natural order of signature letters. These exceptions will be discussed in detail under Collation below.

The issues were sewn together using the standard stitch at the time for "binding" issues of journals, which leaves three small holes centered near the edge of each leaf and plate. This is commonly referred to as "stab sewing."

The page size of the *Arcana*, bound and trimmed, is usually about 14.6 cm wide and 23.3 cm high. From the surviving unbound issues the issue size appears to have been about 15.6 cm wide and 25 cm high. This is the size of a large octavo, but it was printed as a small quarto with eight pages (four leaves) in a signature. The Prospectus on the back cover of the first issue (fig. 2 herein) states that "the purchaser may use his judgement and fancy in selecting such numbers as contain subjects suitable to his taste, where it may not be desirable to purchase the whole; for which purpose the pages are not folio'd [= numbered], in order that Conchology, Botany, Entomology, &c. may be bound up distinctly."

TITLE PAGE—INTRODUCTION

The title page of the *Arcana* is composed of a number of sizes and styles of type. No other title page was ever issued. One copy at hand has what appears to be a Volume II title page and was so taken by this writer for several years. However, examination at magnification reveals that it is a skillfully altered Volume I title page.

The Prospectus on the back cover of number I states that "Twelve numbers will form a volume; at the conclusion of which will be given a Table of Contents . . ." As shown above, this planned format was changed and Volume I is now known to consist of numbers I–XVI. This places the last number of Volume I in 1811 and accounts for that date being shown on the title page. On the cover of number XVIII, the only known cover from Volume II, the "Part" number is replaced with "Vol. II." There is no evidence that the promised "Table of Contents" was ever issued.

The title is followed by a two-page introduction which Perry started by stating that "the subject matter will chiefly consist of such objects in Natural History as have not hitherto been published" but later remarked that "it will not consist entirely of rarities seldom to be seen." It is also stated that it is intended to publish as "an Appendix to each Number" extracts from the remarks of "different English travellers [*sic*]."

As Volume I ended with number XVI (April 1811) it is here assumed that the title page and the introduction are the "missing" Signature pages B_5–B_8 in number XVII. These pages have not been recognized in any sets. This supposition is reinforced by the fact that sets that are "complete" for a period of less than 16 months (e.g., University of Hawaii, Harvard and Smithsonian) lack the title page and the introduction.

DEDICATION

An entire page is used for the dedication. It was issued as a single sheet sewn into number I, as shown by stitch marks on the Oxford University copy. The dedication is to J. C. Lettsom. This was Dr. John Coakley Lettsom (1744–1815), a noted surgeon and philanthropist. Among many other notable acts it was a gift from him that helped make possible the first natural history collection at Harvard University. Unfortunately, he seems to be remembered more for a bit of doggerel that he wrote that does not reflect his attitude toward medicine but is an exceptional play upon his name. It is repeated here and is best when read aloud:

When people's ill, they comes to I,
I physics, bleeds, and sweats 'em;
Sometimes they live, sometimes they die.
What's that to I? I lets 'em.

One of the shells in the first number, *Trochus apiaria*, is drawn "from a specimen in the Museum of Dr. Lettsom". He is not mentioned elsewhere in the *Arcana*, although nine figures in the *Conchology* were drawn from his specimens.

COVERS

This is the first report of the existence of covers for the *Arcana*, although they were certainly to be expected. The Oxford University Museum Library has an unbound partial set of the *Arcana* in original wrappers. These were found during a computer search in which only "Arcana" was entered. This set had not been identified in the Oxford library catalogue as being by Perry as there is no mention of an author on the covers and the cover title differs from that on the title page. The incomplete Oxford set contains numbers I–II, IV–IX, XI–XII and XVIII, all with covers; number III is represented by covers only. No other covers are known for the *Arcana*. They were originally a bright yellow paper that is now very dark and brittle.

Beginning with number IV in April 1810, the date of issue appeared on the cover, following the issue number. Also with that issue, the title was reduced to *Arcana, / or / Museum of Nature.* Between the shortened title and the ornamentation a rectangle was inserted within which the contents of the present number were listed. Earlier issues had listed on the back cover only the contents of the previous and the next numbers and not those of the current number. Lists of contents of previous issues appeared only in numbers II, III and IV. The "contents" consisted of the names of the animal figured and of the artist. At times additional comments were added concerning rarity, the location of specimens and other matters. No mention was ever made in the contents of any of the narrative sections that were included in most issues, but these are listed in the Collation herein.

Beginning with number IX for September 1810, the issue number was preceded by "Part II." This is explained on the back cover of number VIII, where it is stated that ". . . the Volumes of this work will be divided into two Parts, each containing Eight Numbers. Part the First in boards, price £1 may be had of the Publisher." No copies of this first part bound together are known to exist, but it must be remembered that there is probably no author shown and the title differs from that of the *Arcana*, which although rare, may be located under its "final" title in many references. This same cover also states that the recommended arrangement of the subjects will be given in the next number. The recommendation was for "placing each subject (in the binding) under its systematic divisions, as Zoology, Conchology, &c. the exact placing of each object, in respect to priority, being left to their own choice." This explains why some copies are bound by subject or otherwise out of issued order as shown herein under Copies.

Advertisements appeared in three issues, and the back cover of number II is reproduced here (fig. 3). Two-thirds of the back cover of number III was devoted to an ad for a ladies' depilatory. The final advertisement on all available covers is in number IV, the back cover being devoted to an "Exhibition of Velvet Painting" in which it is announced that the artist "undertakes to perfect Ladies in four lessons, who are totally unacquainted with drawing." This seems an unreasonable promise until it is realized that the art involved is probably theorem painting, indicated by his mention that the paintings on exhibition "consist of Window Curtains, Sofas, Chairs, Footstools, Panels, Screens; and the Compositions are Figures, Landscapes, Flowers, Fruit, and various Ornamental Groups." It is not surprising that there are no further advertisements, as the circulation of the *Arcana* must have been very limited. On the back covers of most issues there are paragraphs of appreciation to correspondents and to those who made material available for the plates.

As mentioned at the end of the Collation, the typesetting in the text of the *Arcana* is rather good. The covers, however, contain many misspellings. On the back cover of number II the misspelling of Platypus as "Plattypas" immediately catches the eye as do "Hippocumpus" on number V and "Opussum" on number VIII. In an advertisement on the back cover of number IV, two obvious errors in wording were corrected in pen, almost surely before issuance. In addition to having his name spelled correctly, the artist Whichelo is also shown as "Whichello" and "Whicello". Cruickshank, whose name is discussed under Artists, is spelled as "Cruickshanks" and "Cruikshanks". Other errors, though more in the nature of "changes", abound. In number IX it is announced that a "rare humming bird" will appear in number X, but no hummingbird ever appeared in the *Arcana*. It was replaced in that issue with two species of *Papilio*. Also, a "*Conus dispartus*" was announced for number X but when it appeared the *Conus* was described as *Conus particolor* and the name *dispartus* was not mentioned.

It seems worthy of note that neither the title page nor the covers show George Perry as author, though responsibility for the *Arcana* has never been in doubt. Within the *Arcana*, Perry refers to species in the *Conchology* as "recently described and established by the Editor of this work." However, Perry's name appears as author only on the dedication page and on the "General remarks of the forms and analogy of the toucan, parrot, and eagle" in number XIII.

Callery's (1981) paper on the evolution and importance of part covers, although based primarily on botanical literature, is equally applicable to zoology. She pointed out the various data that can be gleaned from part covers (*e.g.* an interim title, schedule of publication, price, where it may be purchased and other items) that are not to be found in the finished work (or in this case, unfinished). The *Arcana* part covers have certainly provided many details that would otherwise be unknown.

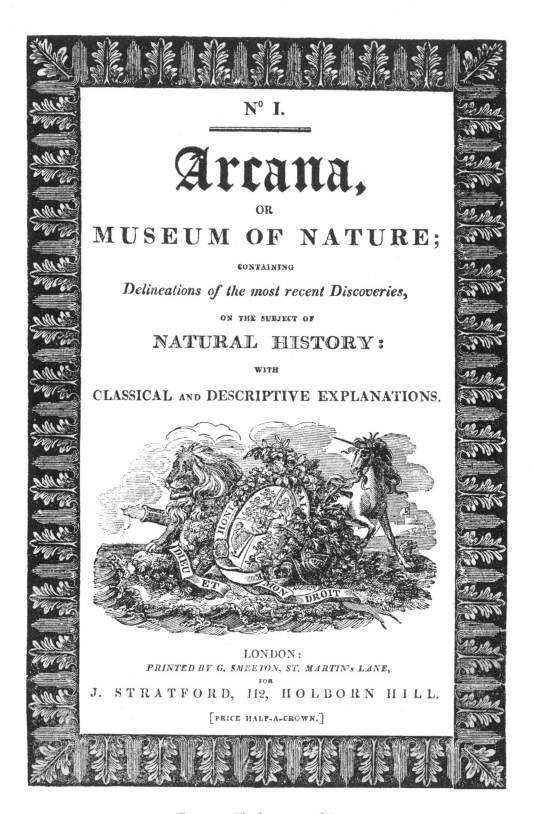

FIGURE I. *The front cover of* No. I

PROSPECTUS.

THIS Work is intended to explain in a pleasing and instructive manner, the most beautiful and interesting Objects of Nature, as they occur in the Families of Animal Life; each number will contain four plates, *faithfully coloured*, with the classical Descriptions suitable to each, according to the most accepted Arrangements, interspersed with Extracts from Narratives of celebrated Travellers. Twelve numbers will form a volume; at the conclusion of which will be given a Table of Contents; and as every number will be complete of itself, the purchaser may use his judgement and fancy in selecting such numbers as contain subjects suitable to his taste, where it may not be desirable to purchase the whole; for which purpose the pages are not folio'd, in order that Conchology, Botany, Entomology, &c. may be bound up distinctly.

All abstruse Terms of Expression will be avoided, or clearly explained, and reduced to an English dress, and made intelligible to the reader.

The Drawings will be executed *from Nature* by PERRY, WHICHELLO, CRUICKSHANKS, and other eminent Artists, who have made the observation of such subjects their peculiar study; and in the descriptive part the Editors will attempt to form a greater accordance of System than has been hitherto observed.

Of the Engravings it is necessary to remark, that Accuracy and Neatness will be the principal characteristics, the zeal of the Artist, Mr. BUSBY, will naturally lead him to emulate the Beauty of the Drawings.

Of the Expense of such a work the Editors are fully aware: but conscious at the same time of the liberal Patronage which has heretofore been generously offered by all Ranks of Society to the Promotion of real Science, they have boldly ventured forth upon the Ocean of Experiment, hoping that their zeal will obtain a lasting and sure reward.

The Knowledge of Nature is the most interesting of all Sciences cultivated by Man, being that upon which all the comforts and satisfaction of life chiefly depend, as affecting more strongly his situation and happiness than any other circumstance; from hence arises the great avidity with which works of Natural History are desired and appreciated.

FIGURE 2. *The back cover of* NO. 1

[9]

A CHEAP OFFER.

A MUSEUM.

GENTLEMEN desirous of forming a Collection of Natural Curiosities, may embrace the opportunity of purchasing 50 Varieties for a Guinea, or 100 for £2. in Shells, Spars, Fossils, Ores, Plants, &c. by Letter, addressed to L. Z. at Mr. STRATFORD's, Bookseller, Holborn.

CONTENTS OF No. I. ARCANA.

I. The TIGER; drawn by Mr. WHICHELLO, from that beautiful Animal in the late Mr. Pidcock's Menagerie.

II. SHELLS, four Specimens; drawn by Mr. PERRY; from Dr. Lettsom's and Lord Valentia's Collections.

III. The PALM TREE; drawn by Mr. PERRY.

IV. The LANTERN FLIES, (two Species) drawn by Mr. PERRY, from scarce Specimens in the Museums of Mr. Stuchbury and Mr. Smith.

No. III. WILL CONTAIN

I. The PLATTYPAS; drawn by Mr. CRUIKSHANKS; from a beautiful Specimen in Mr. Bullock's Liverpool Museum.

II. The POMACEA MACULATA; drawn by Mr. PERRY, from the same interesting Collection.

III. The NONPAREIL PARROTT; drawn by Mr. WHICHELLO; from the same Museum.

IV. The DOLPHIN; drawn by Mr. WHICHELLO, also from the same Museum.

The Proprietors return their sincere Thanks to those Noblemen and Gentlemen, who have so kindly granted them the Use of their Museums and Drawings; and beg leave to assure them, that the utmost Care will be taken of Drawings, &c. and carefully returned when done with.

Mr. S.'s Collection of Drawings of Fishes, is gratefully acknowledged.

The Hints of Mr. P. of Gravesend, are received; his Communications will always be esteemed a valuable Acquisition.

Communications are thankfully received at the Publisher's, Holborn Hill; or the Printer's, St. Martin's Lane, Charing Cross.

G. Smeeton, Printer, 17, St. Martin's Lane, London.

FIGURE 3. *The back cover of* No. II

COLLATION

This collation shows each issue and its date of publication separately. Under the date the first item is the assigned plate number, followed by the signature letter. The page numbers within the signature are shown as subscript numbers. Only the first page of each signature has a letter imprint, the rest necessarily inferred. After the signature designation the subject is stated, usually with the name on the plate first followed by the name used in the text heading or other information considered helpful.

With the exception of the first and last parts, the plate numbers used herein are the same as those assigned by Mathews and Iredale (1912). Their order for the first part was Tiger, Conchology, Botany and *Fulgora*. The first page of text for the Tiger plate is imprinted A, so it is the first plate. The first page of the Conchology text is imprinted as signature B, so it has to follow the first eight pages. The Tiger text consists of four pages, the Conchology of six, the Botany of four, and the *Fulgora* of two. The order in which they are listed here is the only possible arrangement to fit them into eight page signatures.

Mathews & Iredale (1912) arranged the final issue with the Hyena as Plate 83. However, the Hyena text is imprinted Rr and therefore the Hyena is Plate 81. The copy in the Mollusk Department of The Natural History Museum is bound in the order given by Mathews & Iredale. It is notable that Sherborn, who usually cited signature numbers for Perry's taxa, did not do so for *Papilio, Pinus* or *Cassowara* as numbers could not be determined with signature Rr out of place. The placement of these last three parts has been subjective, but when copies were disbound for reassembly, attached pages were discovered showing that the correct order of Rr is Hyena, *Papilio, Cassowara*, and *Pinus*.

There was no signature C in 1810. A signature C did not appear until the issue for May 1811, when it was issued with a second signature B. This has caused several copies to be unintentionally bound out of order. This second signature B appears to consist of only four pages, as no more than four have been located in any copy for this signature. It is supposed that this missing half-signature consisted of the title page and introduction, as discussed under Title Page below. There are no signatures lettered J, V or W in 1810 and no Jj in 1811. Signatures Nn and Oo lack imprinted signature letters.

The two different arrangements of signatures Ee and Ff will be explained at the end of the collation.

Front matter

Title page dated 1811 – verso blank (see comments under Title Page above)
Dedication page – verso blank – issued with Number i.
Introduction – one leaf printed on both sides (see comments under Title Page above)

Text and plates

1 JANUARY 1810

1	A_{1-4}	Tiger – *Felis tigris* – mammal
2	A_{5-8}	Palm Tree – The Ceroxylon – botany
3	B_{1-3}	Conchology – 4 species of mollusks
	B_{4-6}	Remarks on conchology
4	B_{7-8}	*Fulgora* – 2 species of Lantern Flies – insects

1 FEBRUARY 1810

5	D_{1-4}	Rattle Snake – *Crotalus horridus* – reptile
6	D_{5-6}	*Septa tritonia* – mollusk
7	D_{7-8}	Condor Vulture – bird
	E_{1-2}	Ornithology – continuation of discussion about condors
8	E_3	*Sparus – Sparus bandatus* – fish
	E_{4-8}	Ichthyology – discussion of various fish

1 MARCH 1810

9	F_{1-4}	Nonpariel [sic] Parrot – *Psittacus nonpareil* – bird
10	F_{5-8}	Platypus – *Ornithorinxus paradoxus* – mammal
11	G_{1-4}	Green Parrott [sic] – *Psittacus viridis* – bird
12	G_{5-6}	*Pomacea maculata* – mollusk
	G_{7-8}	An account of the Termites Bellicosus, or White Ants, found in Africa; extracted from Mr. Smeathman's travels in that country.

1 APRIL 1810

13	H_{1-2}	Dolphin – fish
14	H_{3-4}	Vicuna – mammal
15	H_{5-8}	Fossils – 4 species of fossil mollusks
	I_{1-6}	Continuation of remarks on fossils that began at the bottom of H_5
16	I_{7-8}	Gloria Maris – *Conus gloriamaris* – mollusk

1 MAY 1810

17	K_{1-4}	Koalo – The Koala – mammal
18	K_{5-6}	Sea Horse – *Hippocampus foliatus* – fish
	K_{7-8}	Article on ants continued from G_8
	L_{1-4}	Article on ants continued
19	L_{5-6}	Bulimus zebra – mollusk
20	L_{7-8}	Military Macaw – *Ara militaris* – bird

1 JUNE 1810

21	M_{1-4}	Wombach – *Opossum hirsutum* – Wombat – mammal
22	M_{5-6}	New Holland Crane – *Ardea rubicunda* – bird
23	M_{7-8}	*Triplex foliatus* – mollusk
24	N_{1-2}	*Mantis foliatus* – insect
	N_{3-8}	Extract of an Account of a Tour made at the top of the Peake of Teneriffe, by Mr. Glas, in the year 1761

1 JULY 1810

25	O_{1-4}	*Chamaeleo – Chamaeleo pallida* – reptile
26	O_{5-6}	*Stromateus – Stromateus depressus* – fish
27	O_{7-8}	*Dipus muscola* – no name on plate – mammal
28	P_{1-2}	*Scalaria – Scalaria disjuncta* – mollusk (five mollusks on plate)
	P_{3-8}	Extract from Mr. Patterson's Travels in Africa, in the year 1778.

1 AUGUST 1810

29	Q_{1-4}	King Bird, of Paradice [*sic*] – *Paradisea regia* – bird
30	Q_{5-6}	*Voluta pacifica* – mollusk
31	Q_{7-8}	*Papilio demosthenes* – insect
32	R_{1-2}	Opossum Mouse – mammal
	R_{3-8}	An Account of the Religion and Superstitition of the Modern Africans.

1 SEPTEMBER 1810

33	S_{1-4}	Tortoise – *Testudo panama* – reptile
34	S_{5-6}	Club Echinus – *Echinus castaneus* – echinoderm
35	S_{7-8}	*Triplex – Triplex flavicunda & Triplex rubicunda* – mollusks
36	T_{1-2}	Papuan Lory – *Psittacus papuensis* – bird
	T_{3-8}	Extracts from Barrow's Travels in China.

1 OCTOBER 1810

37	U_{1-2}	*Papilio* – 2 species; no names on plate – insects
38	U_{3-4}	*Echinus stellaris* – echinoderm
39	U_{5-6}	*Conus particolar* [*sic*] – *Conus particolor* – mollusk
40	U_{7-8}	Sanglin Monkey – *Sapajus jacchus* -- mammal
	X_{1-2}	Continuation of discussion on Sanglin Monkey and others
	X_{3-8}	Extracts from the Travels in China, from Sir George Staunton's Account.

1 November 1810

41	Y_{1-2}	*Bradypus striatus* – no name on plate – mammal
42	Y_{3-6}	*Monoculithos – Monoculithos gigantea* – trilobite
43	Y_{7-8}	*Bulimus phasianus* – mollusk
44	Z_{1-2}	*Cerithia coerulea* – bird (this plate was omitted by Mathews & Iredale).
	Z_{3-8}	Extracts from Phillip's Account of the Customs and Arts of the Natives of New Holland.

1 December 1810

45	Aa_{1-2}	*Hippocampus – Hippocampus erectus* – fish
46	Aa_{3-4}	*Arcuatus coeruleus* – insect
47	Aa_{5-6}	*Aranea gracilis* – mollusk
48	Aa_{7-8}	*Monoculithos* – 2 species – trilobites
	Bb_{1-8}	Extracts from the Travels of Mr. Mungo Parke in the Interior of Africa, containing an Account of the River Niger.

1 January 1811

49	Cc_{1-4}	Cameleopard – *Camelus camelopardalis* – mammal
50	Cc_{5-6}	Red Phalarope – *Tringa rubra* – bird
51	Cc_{7-8}	*Phalaena – Phalaena vitrea* – insect
52	Dd_{1-2}	*Strombus solitaris* – mollusk
	Dd_{3-8}	General remarks on the forms and analogy of the toucan, parrot, and eagle.

1 February 1811 – original printing

53	Ee_{1-5}	Mountain Cow – *Antelopa montana* – mammal
54	Ee_{6-7}	*Aranea conspicua* – mollusk
55	Ee_8	*Congiopodus percatus* – fish
	Ff_1	Continuation of remarks on *Congiopodus*
56	Ff_{2-3}	Pelican – *Pelicanus africanus* – bird
	Ff_{4-8}	Remarks on the Locusta, or Gryllas, a Native of Arabia Felix; Remarks on the cochineal insect (of South America); Of the lac cochineal insect (of the East Indies); Of the wax-forming cicada.

February 1811 – reissue

53	Ee_{1-4}	Mountain Cow – *Antelopa montana* – mammal
54	Ee_{5-6}	*Aranea conspicua* – mollusk
55	Ee_{7-8}	*Congiopodus percatus* – fish
56	Ff_{1-2}	Pelican – *Pelicanus africanus* – bird

Ff$_{3-8}$ Remarks on the Locusta, or Gryllas, a Native of Arabia Felix; Remarks on the cochineal insect (of South America); Of the lac cochineal insect (of the East Indies); Of the wax-forming cicada.

1 MARCH 1811

57   Gg$_{1-4}$ Lion – *Felis leo* – mammal

58   Gg$_{5-6}$ *Buccinum dilatum* [plate] – *Buccinum orbiculare* [text] – mollusk

59   Gg$_{7-8}$ Black Swan – *Anas niger* – bird

60   Hh$_{1-2}$ *Papilio catenaria* – insect

      Hh$_{3-8}$ Remarks on the different Genera of Insects as elucidated in the Seven Orders of Linnaeus.

1 APRIL 1811

61   Ii$_{1-4}$ Elephant – *Elephas gigas* – mammal

62   Ii$_{5-6}$ *Trochus zebra* – mollusk – [plate is misdated March 1, 1811]

63   Ii$_{7-8}$ Panther – *Felis pantherus* – mammal

64   Kk$_{1-2}$ *Esox niloticus* – fish – [plate is misdated March 1, 1811]

      Kk$_{3-8}$ Remarks on the infusoria, insects, and polypi, discernable only by the microscope.

1 MAY 1811

65   B$_{1-2}$ Leopard – *Felis leopardis* – mammal

66   B$_{3-4}$ *Buccinum distentum* – mollusk – [plate is misdated May 1, 1810]

      B$_{5-8}$ No pages for this half-signature have been located (see comments under Title Page below).

67   C$_{1-2}$ Babyroussa – *Babyrousa quadricornua* – mammal

68   C$_{3-4}$ Guanaco – *Guanaco patagonia* – mammal

      C$_{5-8}$ General Remarks on the King-fisher Tribe.

1 JUNE 1811

69   Ll$_{1-2}$ *Sphynx castaneus* – insect

70   Ll$_{3-4}$ *Lanius aurantius* – no name on plate – bird

71   Ll$_{5-6}$ *Pecten sanguineum* – mollusk

72   Ll$_{7-8}$ Fossil Elk – *Cervus fossilis* – mammal

      Mm$_{1-8}$ Extracts from the Rev. W. Bingley's Animal Biography. Of the Camel Tribe & Of the Arabian Camel.

1 JULY 1811

73   [Nn]$_{1-2}$ Kanguroo [*sic*] – *Dipus tridactylus* – mammal

74   [Nn]$_{3-4}$ *Phalaena fenestra* – insect

75   [Nn]$_{5-6}$ *Strombus divergens* [plate] – *Strombus nigricans* [text] – mollusk
76   [Nn]$_{7-8}$ Zebra – *Equus zebra* – mammal
        [Oo]$_{1-8}$ Remarks on the Natural History of the Buffalo.

1 August 1811

77   Pp$_{1-2}$ Aegyptian Ram – *Ovis aries* – mammal
78   Pp$_{3-4}$ Crown Crane – *Ardea coronata* – bird
79   Pp$_{5-6}$ John Dory – *Zeus faber* – fish
80   Pp$_{7-8}$ *Conus bandatus* – mollusk
        Qq$_{1-8}$ A Dissertation on the Corvus Tribe of Birds of the Crow Kind.

1 September 1811

81   Rr$_{1-2}$ Spotted Hyena – mammal
82   Rr$_{3-4}$ *Papilio volcanica* – insect
83   Rr$_{5-6}$ Cassowary – *Cassowara eximia* – bird
84   Rr$_{7-8}$ *Pinus* – *Pinus rostellaria* – mollusk
        Ss$_{1-8}$ Remarks on the Penguin of the South Seas, and other Marine Birds.

As shown above, there were two printings of the text for the February 1811 issue (signatures Ee and Ff). Evidently the compositor and/or the printer erred in allowing the Mountain Cow text (Ee$_{1-5}$) to end on an odd-numbered page. This meant that page Ee$_5$ was printed on the recto of the first page of the *Aranea* text. The *Aranea* text consisted of only two pages, so its final page was on the recto of the "Pelican" text and so on. As the text for all plates to that date started on a recto (odd numbered) page, allowing the plate to face the first page of the appropriate text, these signatures were reprinted to make them uniform and to enable the work to be divided by subject as promised in the Prospectus.

It may be speculated that this error escaped Perry's attention at first as he was busy with the production of his *Conchology*, which by January of 1811 had been in preparation for almost a year and was obviously causing him some problems. The introduction to the *Conchology* is dated January 1811, but it did not actually appear until April, the same month in which these two signatures would have been printed.

Several copies of the *Arcana* include both sets of text, but as most have only one version the differences will be briefly described. It is obvious that the type had been disturbed or possibly disassembled completely and required resetting, as the first two pages were not reworded but errors were introduced. On Ee$_1$ the word "Poets" in the middle of the page lost its capitalization in the reprint, and in the last line the word "also" was corrupted to "alo." There is another error introduced in the middle of Ee$_2$ with "its" changed to "iss." A five-line paragraph in the middle of Ee$_3$ containing general remarks about antelopes was dropped, and seven lines from the top of Ee$_4$ were added to new page Ee_3. On page Ee_5 the last nine and a half lines are moved to become part of the last paragraph on new page Ee_4, being attached to the upper paragraph of old Ee_6, the last two paragraphs on that page being deleted. There are two minor changes in capitalization, and there is an addition to the text. After the statement about "animals . . . preserved alive in the Tower" the following is inserted: "(from whence the accompanying representation was delineated)." The text for the Pelican is unchanged, but the top sentence on old Ff_3 was moved to the bottom of Ff_1, to which the signature letters were also added. With the rearrangement moving Pelican up to Ff_{1-2} without any additional text, six pages were left instead of five for the narrative on Locusts. No text was added, but the type was entirely reset in a larger font size. Also, the running title at the top of each page was changed from "Entomology" to "Remarks on the Locusta", etc. Even then the final page is only one-half occupied.

PRICE

The price of each issue was "half-a-crown," a unit of currency no longer in use that does not convert easily to the decimal system as it is one-eighth of a Pound. There was no extra charge for the boards in the offer of the first eight numbers for £1, so that half-volume can be used as a basis for comparison. It is impossible to have an absolute equivalence between currencies separated by almost two centuries. Remembering that the first eight numbers contained 32 colored plates it can be noted that Mawe's *The Linnaean System of Conchology* with 37 colored plates was published in 1823 at £2 1 2s 6d. It appears that the *Arcana* was not overpriced, but in addition to being a serial publication it covered all of natural history, so the comparison is far from exact. In 1810–11 the sum of £1 had considerable purchasing power.

RARITY

Mathews & Iredale (1912: 7) knew of only four copies of the *Arcana:* "one in the Natural History Museum, South Kensington; one in the library of the Zoological Society, London; one in Sweden [Stockholm University Library], and the fourth in our private library at Watford." This fourth copy was donated to the National Library of Australia by Mathews as part of his ornithological library. The other three are still at the locations cited. Pilsbry (1927: 64) added the copy at the Academy of Natural Sciences of Philadelphia as the fifth known, with the remark that "It is a diverting and occasionally quite thrilling book." Whittell (1954: 583) did not have a copy available when he wrote his history and bibliography of Australian ornithology, as he copied from Mathews & Iredale and thus also listed thirteen plates of birds instead of the correct fourteen.

Attesting to its rarity is a comment made by the late C. Kirke Swann, whose family owned the famous natural history rare book firm Wheldon & Wesley Ltd. He wrote:

A mention of George Perry's Arcana; or Museum of natural history (1810–11) always reminds me of an afternoon about 1928. A catalogue from the firm of Backus in Leicester arrived in the post and I was told to go home, pack a bag, and catch the evening train to Leicester. I was to be on Backus' doorstep when the shop opened next morning and get the Perry. I did as I was told and brought it home. It cost 7s. 6d., but even with the expenses it was cheap, as we sold it for £28—a lot of money in those days.—Swann, 1972: 121

The present author previously estimated the number of copies as eight (*in* Kohn 1986: 3) but extensive searches, in which Scott Jordan was very active, have now identified 32. Most are unfortunately incomplete.

The copies located to date are as follows:

United Kingdom [8]

- The Natural History Museum, London
  Mollusk Department   complete
  Zoology Library, Rare Books   complete
  Zoological Museum, Tring: Copy 1   complete (bound by subject)
  Zoological Museum, Tring: Copy 2   lacks plates 54 & 84
- Zoological Society of London   lacks parts 18 & 21
- Oxford University Museum Library   parts 1–9, 11, 12, 18 plus some duplicates. A very important unbound set with the only known covers
- Natural History Society of   complete (1 extra plate)
  Northumbria, Newcastle-upon-
  Tyne
- Location unknown; sold at auction   complete
  2006

Sweden [1]

- Stockholm University Library   complete; 2 signatures bound out of order

Australia [11]

- Australian Museum, Sydney   complete
- University of Melbourne   lacks 36 plates (has one duplicate)

- Library of the Museum of Victoria, Melbourne — text and plates bound separately; lacks 5 plates and about 36 text pages
- State Library of South Australia, Adelaide — complete; bound and arranged by subject
- State Library of New South Wales, Sydney — parts 1-12 only
- State Library of Western Australia, Perth — lacks 5 plates
- Western Australian Museum, Perth — lacks 29 plates
- Monash University, Victoria — lacks plates 60 & 82
- CSIRO Black Mountain Library, Canberra — collation not available; stated to have 48 color plates, but also listed as having 13 parts
- National Library of Australia, Canberra:
  Copy 1 — complete, bound arranged by subject
  Copy 2 — complete, bound with part 17 out of order

Canada [1]

- McGill University, Quebec — lacks 12 plates

United States [11]

- Academy of Natural Sciences, Philadelphia — complete
- American Museum of Natural History, New York — lacks parts 20 & 21 and also pls. 9 & 33
- Cornell University, Ithaca, NY — complete
- Harvard University, Cambridge — parts 1–12 only
- Smithsonian Institution, Washington — parts 1–13 only
- University of Hawaii, Manoa — parts 1–9 only
- Scott Jordan, La Habra Heights, CA
  Copy 1 — complete
  Copy 2 — lacks 14 plates
- R. E. Petit, North Myrtle Beach, SC
  Copy 1 — complete (except for 16 pp. of narratives)
  Copy 2 — lacks 3 plates
  Copy 3 — lacks 32 plates

Of the 32 copies listed above, only thirteen contain all 84 plates. Copies are considered complete if they contain either of the two printings of Signatures Ee and Ff. Collations of all are not available but it is assumed that copies lacking plates also lack the corresponding text. Based on extensive searches of sales catalogues and information supplied by rare book dealers, it is estimated that two more copies may be in private hands, one each in Australia and the United States. Other unrecorded copies might exist in institutions or in private collections. The known copies not traced might also have changed hands and thus actually be in the above tally.

In August 1956 an incomplete copy containing 52 colored plates was sold by Wheldon & Wesley Ltd. to the Commonwealth Parliamentary Library, the progenitor of the National Library of Australia. Mr. Paul Livingston of the NLA contacted a colleague at the Parliamentary Library who confirmed that they do not hold a copy. As can be seen from the above list, no Australian copy fitting that description has been located. It is easy to notice that it is always the later parts that are lacking.

PLATES, ARTISTS AND ENGRAVERS

The 84 plates of the *Arcana* are reproduced herein with numbers matching those given in the Collation. The derision heaped upon the illustrations in Perry's *Conchology* (see Petit 2003) cannot be applied to the *Arcana*. As can be seen, with few exceptions the plates are well drawn and colored and are equal in quality to other plates of the time. The illustrations were the work of five different artists, with the majority drawn by George Perry himself. He executed 47 in all, including the mollusk, trilobite, and insect plates, the one botany plate, and others. T. L. [Thomas Lord] Busby, who engraved all of the *Arcana* plates, also drew 21 of them (twelve mammals, eight birds and one echinoid). Nine plates were drawn by J. C. Whichelo, who has not been identified with certainty as a number of artists in the Whichelo family flourished at the time. Five mammals were drawn by "Cruikshanks," who cannot be further identified as there is no initial before the name. No engraver or artist named Cruikshanks (with a final 's') can be located in London in 1810-1811. There were three well-known artists named Cruikshank (without a final 's'; it also appears as Cruickshanks on Plate 40): the father, Isaac (1756–1811; the last plate attributed to Cruikshanks was for 1 October 1810) and two sons, Isaac Robert (1789–1856) and George (1792–1878). The brothers became famous primarily through their engravings. A search of the on-line version of the Oxford Dictionary of National Biography, (Patten 2004–5a) for simply "Cruickshank" turns up a list of the family members and following George's name is the notation "also known as Cruickshanks." That information does not appear in the article itself. It is stated there that "many prints were collaborative efforts" and that may account for the "s." However, on the covers these plates are attributed to "Mr. Cruikshanks."

This accounts for all but the two plates that were drawn by C. Hirst, about whom nothing is known. Both Hirst plates are of birds, and are stated in the text to be based on drawings sent by Mr. Priestnall of Stockport. The only plate that does not show the name of the engraver is Plate 70, but it most probably was also the work of Busby.

In number XII, which had been scheduled for two figures by Busby but instead contained four by Perry, this announcement appeared on the back cover: "Owing to the Indisposition of Mr. Busby, the Editors humbly solicit their Friends to overlook the alterations in the Subjects of the present Number of the Arcana, which have been unavoidably caused by the above circumstance." There are other instances on the available covers indicating that Busby may have had a predisposition to indisposition, but at least he got all the plates engraved even if he did not draw some as planned.

One incomplete copy of the *Arcana* at hand includes seven plates that are uncolored. That uncolored plates exist would not be strange except that two of these have Whatman watermarks on which the last half of the date appears as "21" [= 1821], or ten years after the plates were first printed. Also, the name of the publisher and the publication date have been removed from all seven. In another copy there is an uncolored Plate 5 but the publisher's name has not been deleted and Plate 49 in the McGill University copy is uncolored. No other uncolored plates have been located, which is even stranger, and brings questions for which no answers can be provided: Why are no more of these plain plates known? In what form were they distributed (the corresponding text is not present in the copy containing the uncolored plates)? Why was the publisher's name removed? The answer to the last question is easy to guess, but who actually published them? In the copy of the *Arcana* with the seven plain plates there are also two colored ones from which the publishers' name has been deleted (Plates 7 and 8). The uncolored plates are from three different Parts: 53, 55 & 56; 73 & 75, and 82 & 84. These do not appear to have been produced to fill in for missing color plates, as the corresponding text is lacking. Also, this copy does not contain either plates or text for the other parts of the respective issues. It is obvious that the original engravings were used at a later date, but why have they not been noted before, why have so few been found, and why didn't this late (re)production increase the number of known copies? Of all of the odd phenomena associated with Perry's works, this is perhaps the strangest.

The *Arcana* plates are unusual for their time in that some were printed in a color other than black. Some were printed in the predominant color of the animal figured. As an example, the Koala plate is printed in brown with the other colors and shading added by hand as usual. This results in the hairs extending past the silhouette of the Koala being brown, rather than either being black or having a brown-wash background.

Arcana plates are often reproduced in books on the history and the natural history of Australia (e.g. Moyal, 1986; Finney, 1984), with the Koala being extremely popular. The Koala plate now appears on numerous web sites and is featured on the Australian Museum Research Library home page. Unfortunately, the great majority of references to it are misdated as 1811 instead of 1810. It was reproduced by both Moyal (black and white, enlarged) and Finney (color, reduced). Both authors also reproduced the wombat plate (Plate 21) in reduced size, Moyal in black and white and Finney in color. Finney's book also has color reproductions of the "Green Parrot" (Plate 11) in reduced size and the "Seahorse" (Plate 18) in original size. All of these plates date from 1810 although they are shown as 1811 by both authors. However, in his text Moyal mentions the "*Arcana* in 1810."

WATERMARKS

Only a few copies have been examined for watermarks. Most of the *Arcana* is printed on unwatermarked paper. A few leaves of text are on paper with only the date as a watermark and no indication of the papermaker or mill. Some 1808 watermarks are present on leaves printed from January through April 1810; the title page is watermarked 1809, as are two text pages printed in February and August 1810. An 1810 watermark is on a leaf printed in August 1811. Watermarks were found on only two other signatures of text, both printed in June 1811. This watermark, split on the fold as expected in a quarto work, reads "Buttanshaw 1806" with the name on one line and the date below. No other Buttanshaw paper was found in the copies examined.

At least three different papermakers furnished paper for the plates. The watermarks are all split but it is not difficult to determine the full name of the maker, although the right hand side of the mark is necessary to determine the year. The first watermarked plates noted were published in January 1810, watermarked "Edmeads & Pine 1808". No plates with that watermark appear after July 1810. Plates watermarked "Ruse & Turners 1810" were published from August 1810 to June 1811. Plates published in January, February and March 1811 have "J Whatman 1810" watermarks. One February 1810 plate also has a "J Whatman" watermark but only the left half is present and it thus cannot be dated.

For a note about 1821 watermarks on uncolored plates, see under "Plates, artists and engravers" above.

PUBLISHER, PRINTER AND "INDEX"

The *Arcana* was published by James Stratford of 112 Holborn Hill, and his name appears at the bottom of the plates. He was a printer, bookseller and stationer. Stratford was in business at that same address until at least 1816 (Exeter Working Papers, 2005).

In his major work on the literature of vertebrate zoology, Casey Wood (1931: 517) consulted the incomplete *Arcana* at McGill University Library, referring to it as a "rare treatise." Wood cited the British Museum (Natural History) Catalogue listing of 84 plates and then stated: "The present volume is 8vo and contains 72 plates only, one of which is uncolored. A MS note states that 'this volume contains all the numbers that were published of this work; the Publisher failed'." The author of that note is not known but it is still present in the volume and it is not in Casey Wood's handwriting. It would be interesting to know the source of this misinformation and Wood's reason for repeating it, as he knew the British Museum had a copy with 84 plates.

The McGill copy was also stated to lack the index, but none was ever issued. There is a handwritten index that Wood evidently thought was copied from a printed one rather than created by a previous owner. As shown earlier, a Table of Contents was promised but evidently never issued.

The *Arcana* was printed by George Smeeton of 17 St. Martin's Lane. Smeeton printed and published numerous works illustrated by the Cruikshank family (Patten, 2004–05b). He remained at that address until at least 1828.

SOURCES

Perry obtained specimens to figure from a number of places. Bullock's Museum was cited as the source of 25 figures, followed by Polito's Menagerie with eleven. Two other menageries were credited: Mr. Kendrick's, with three figures, and Mr. Pidcock's with one. Of the thirteen individuals and the British Museum, only two are mentioned twice and the others only once. In number III (March 1810), in the "Acknowledgements to Correspondents" on the inside back cover, there is an acknowledgement of "the generous Assistance of Mr. Bullock, for the Use of those admirable Resources of Specimens so ably arranged in his extensive Museum at Piccadilly." The following paragraph states that "The same Regard is also due to Mr. Polito, the late Mr. Pidcock, and Mr. Kendrick . . ."

The last plate figuring a specimen from Bullock's was Plate 54, published February 1811. The first plate figuring an animal from Polito's was issued in October 1810, and the second in February 1811. After that date Bullock's is not mentioned. What this signifies, if anything, is not known. It has been noted (Petit, 2003: 10) that Bullock did not use any of Perry's *Arcana* plates in his 1812 *Companion,* although they are the same size and type, and sometimes even of the same species. Bullock did not mention Perry in any way or use any of his names but he proudly listed: "The small Bell Glass, number 4, contains several articles which were once the property of the celebrated Sir Charles Linnaeus; a specimen of whose writing is likewise inclosed [*sic*]. Presented by Dr. Smith, President of the Linnaean [*sic*] Society" (Bullock, 1812: 123). These items were previously listed by Bullock in his unillustrated *Companion* of 1810 (eighth edition, p. 77). There are only three articles listed and it will be of interest to some to see the list (italics as in original):

1. *Isis versicolor,* Many-coloured Isis Coral, orange var. New S. Wales
2. *Physeter* [*sic;* actually a mammal genus] perspectivus, Perspective or Staircase-shell. China. Trochus perspectivus Linn.
3. *Scala grandis,* Great or true Wentletrap. Sumatra. Turbo scalaris Linn.

This list is followed by the attribution to Smith and Linnaeus given above. This eighth edition of Bullock's *Companion* is the first to list these Linnaean specimens and was also the edition with the most space devoted to shells. The ninth edition, also published in 1810, does not describe these articles but lists them as in the 1812 twelfth edition. The Linnaean shells are not mentioned in the tenth edition of 1811. As a matter of possible interest, no eleventh edition is known.

It is easy to conjecture that Bullock and Perry discontinued their association due to the influence of Smith, but that can only be a guess. Bullock was elected a Fellow of the Linnean Society of London on 11 June 1810.

Although considerable time and effort has been expended in an effort to locate specimens named or figured by Perry, none have been found to date.

TEXT

Perry's promised extracts from the remarks of various English travelers appear in several issues. They are mostly standard travelogues of the time, some more "scientific" than others. These are listed in the Collation, together with Perry's papers that do not directly refer to plates.

In addition, Perry often wrote at length about various subjects. His comments must be considered in the context of zoological and geological science in the first decade of the nineteenth century. Remember that it was only a few decades earlier that the question whether or not fossils were the remains of once living creatures or were only "formed stones" was being debated. The rigid structure and arrangement of the Linnaean system had been challenged by very few in Britain. Perry was well read and often quoted Lamarck and others whose works were not yet accepted by his fellow countrymen. In the very first issue he stated that certain "multivalvae" of Linnaeus such as chitons "ought properly to be separated" from the "Testacea." Later, in the August 1810 issue (signature Q_6), Perry once again referred to the Linnaean system with a comment about "the modern French writers, particularly Monsieur Bruguiere [*sic* = Bruguière] and Lamarck, whose system is far more perfect and complete, although by no means opposite to that of the great Linnaeus."

In the first issue there is also a discussion of the manner of growth of gastropods, listing various arguments about the mechanism used. Unfortunately, Perry adopts the wrong hypothesis, stating that "it seems more probable that that the Shell has an internal Power of Growth or Expansion, which exists from its beginning or birth, and adapts itself by a general Expansion to the Size of the Animal" [capitalization as printed]. At the end of this discussion Perry mentioned that most shells "revolve spirally from the left to the right," and listed some genera that are exceptions.

In a discussion on fossil mollusks in the April 1810 issue, Perry speculated that a comet passing close to the earth would affect the inclination of the earth's axis causing shifting of waters sufficient to flood various parts of the earth, thus accounting for the fossil formations previously attributed to the Deluge. This might seem to be an original idea to those of us unfamiliar with the history of science, and in fact was so taken by Finney (1984: 137). However, in another discourse in the June 1811 issue, Perry mentioned this again and credited the theory to "Whiston" (William Whiston, 1667–1752), whose New Theory of the Earth was published in 1696. Prof. Hugh Torrens has advised that this idea resurfaced regularly and was reiterated by Farey in 1808 (see Torrens, 2002: vi: 12). Perry was quite taken with the fossil fauna and devoted two plates of the *Arcana* to fossil mollusks of the Paris Basin and the Barton beds of Hampshire (see Petit & Le Renard, 1990) as well as two plates of English trilobites and one of an "Irish Elk."

Perry had unusual ideas about animals. His comments on the Koala are often repeated and were quoted, at least in part, by Finney (1984: 137), Moyal (1986: 42) and on several web sites. In brief, Perry commented:

> Whether we consider the uncouth and remarkable form of its body, which is particularly awkward and unweildy [*sic*], or its strange physiognomy and manner of living, we are at a loss to imagine for what particular scale of usefulness or happiness such an animal could by the great Author of Nature possibly be destined. . . . As Nature however provides nothing in vain, we may suppose that even these torpid, senseless creatures are wisely intended to fill up one of the great links of the chain of animated nature, and to shew forth the extensive variety of the created beings which God has, in his wisdom, constructed.

Perry would not have believed that the Koala would become not only the national symbol of Australia but a favorite attraction at zoos worldwide.

The typesetting of the text is generally good. The Academy of Natural Sciences copy has a typographical error in the sixth line on signature U_2, where *Papilio* is misspelled as *Papilis*. This was apparently corrected quickly as the misspelling has been noted only in that copy, though admittedly an extensive search was not made for this correction. An error of a different type is found in signature Ii, where the correct running title on Ii_5, Conchology, becomes Ornithology on Ii_6.

TAXA

In each group, for newly introduced nomina, the name will be followed by: Perry, year, (month), signature number of page where name first appears and plate number. For taxa not newly introduced, the name will be followed by: *Arcana,* year, page range and plate number. Names introduced by Perry and in current usage are in bold type.

Insecta

There are ten plates of various orders of insects in the Arcana, illustrating a total of twelve species. Two of Perry's species names are in current usage. He introduced three genus group names, none of which are currently in use. One was introduced as a "Division" and is simply listed here, as it has never been listed or treated as a valid name elsewhere.

Arcuata Perry, 1811 (1 January), signature Cc$_7$, Pl. 51; name appears on Cc$_7$.

Type species, by monotypy, *Arcuata vitrea* Perry, 1811. As shown by Fletcher & Nye (1982: 19), Perry's species *A. vitrea* is an unnecessary replacement name for *Phalaena atlas* Linnaeus, 1758 and therefore is a junior objective synonym of that species. Consequently, the genus name *Arcuata* is a junior objective synonym of *Attacus* Linnaeus, 1767. *Arcuata* was not listed as a genus group name by either Sherborn or Neave, although Sherborn listed the species name as *Arcuata vitrea* Perry. The name *Arcuata* appears only in the text heading on signature Cc$_7$.

Arcuatus Perry, 1810 (1 December), signature Aa$_3$, Pl. 46.

Type species, by monotypy, *Arcuatus coeruleus* Perry, 1810. The species name appears only on the plate. In the text heading *Arcuatus* is shown as a Division, but is clearly intended as a genus name as shown on the plate, and it was so accepted by Mathews & Iredale, Sherborn and Neave. Perry's usage of nomenclatural terms has been addressed by Petit & Le Renard (1990). Mathews & Iredale (1912: 16) discussed Perry's proposal for a new classification of the Papilionidae, listing his six divisions as "Arcuati [*sic*], Orbati, Caudati, Excelsi, Cuspidati and Muscarii." This is correct except for "Arcuati", which Perry did not use in that form and which can only be inferred from *Arcuatus*. *Arcuatus* is considered to be a junior subjective synonym of *Morpho* Fabricius, 1807.

Orbatus Perry, 1811 (1 September), signature Rr$_3$.

Type species, by monotypy, *Papilio volcanica* Perry, 1811. The text-heading for this species is formatted thus: *Genus* – PAPILIO. *Species* – PAPILIO VOLCANICA. *Division* – ORBATUS. This has to be accepted as a genus group name under Article 10.4, although it has never appeared as such in the literature. It is included here only for completeness. In this compilation the name is intended as a listing in the sense of Article 23.9.6 and not as a senior synonym of any taxon. It will almost certainly be properly declared a *nomen oblitum* by concerned lepidopterists in an appropriate publication if it is found to conflict with any current usage.

candelaria (Fulgora) – Arcana, signature B$_8$, Pl. 4, fig. 2.

= *Pyrops candelarius* (Linnaeus, 1758), the Hong Kong Lantern Fly. Not attributed by Perry, who stated it to be a native of China.

castaneus (Sphynx) Perry, 1811 (1 June), signature Ll$_1$, Pl. 69.

= *Coequosa triangularis* (Donovan, 1805) *fide* Dr. Kitching. This species is from Australia and Tasmania.

catenarius *(Arcuatus)* Perry, 1811 (1 March), signature Hh$_1$, Pl. 60.

= *Morpho epistrophus catenaria* (Perry, 1811). Although on the plate legend as *Papilio catenaria,* it is *Arcuatus catenarius* in the text heading. A Brazilian subspecies.

coeruleus *(Arcuatus)* Perry, 1810 (1 December), signature Aa$_{3-4}$, Pl. 46; name only on plate.

= *Morpho menelaus coerulea* (Perry, 1810), a subspecies of *Morpho menelaus* (Linnaeus, 1758). Perry's species name appears only on the plate, where it is rendered with the "oe" as a ligature that is clearly "oe" and not "ae." Later, in the discussion under *Arcuatus catenarius,* Perry (1811: signature Hh$_1$) spelled this as *caeruleus* with the "ae" as a ligature. For discussion of the genus name, see above under *Arcuatus.* The subspecies occurs in southeastern Brazil.

corollaria (Phalaena) Perry, 1810 (1 October), signature U$_2$, Pl. 37, fig. 2.

= *Automeris io* (Fabricius, 1775) *fide* Dr. Kitching. Listed by Mathews & Iredale (1912: 16) without comment, but not listed by Sherborn. Perry correctly gave the locality as North America.

demosthenes (Papilio) Perry, 1810 (1 August), signature Q$_7$, Pl. 31.

= *Caligo beltrao* (Illiger, 1801). The locality given by Perry was "the Brazils." The species occurs in southeastern Brazil.

fenestra (Phalaena) Perry, 1811 (1 July), signature [Nn]$_3$, Pl. 74.

= *Antheraea (Telea) polyphemus* (Cramer, 1775) *fide* Dr. Kitching. Incorrectly listed by Sherborn (1926: 2353) as *P. fenestrata.* No locality was given by Perry. The species ranges throughout North America.

foliaceus (Mantis) – Arcana, 1810, signature N$_1$

= *Mantis foliatus, q.v.*

foliatus (Mantis) – Arcana, 1810, signature N$_{1-2}$, Pl. 24.

= *Phyllium siccifolium* (Linnaeus, 1758). The specific name is *foliaceus* in the text heading, the only place it appears, and *foliatus* on the plate. Mathews & Iredale (1912: 15) listed it as *M. foliaceus* without comment, but Sherborn (1926: 2476) listed it as *M. foliatus.* This is certainly a misidentification by Perry of *Mantis foliata* Lichtenstein, 1802, now placed in the genus *Vates.* In his discussion Perry referred to Lichtenstein's work "in the Sixth Volume of the Linnean Transactions" where reference is made to *M. foliatus* Lichtenstein (*fide* Sherborn 1924: 2476), so it is obvious that Perry did not intend introducing this as a new name but cited it as of Lichtenstein. Perry did not give a locality, but this is an Asian species.

phillis (Papilio) – Arcana, signature U$_2$, Pl. 37, fig. 1.

= *Heliconius erato phyllis* (Fabricius, 1775). *Papilio phillis* was attributed to Fabricius by Perry, but he misspelled the species name. Dr. Brian Pitkin notes that the red bands on the forewings appear to be the wrong color, appearing slightly bluish as if originally colored differently and then painted over. Comparison of two copies at hand shows one colored as described by Dr. Pitkin but the other with red bands as described in the text. However, on that "normal" plate there are smears of blue paint, the only plate in the entire volume so marked. Two other copies also have the red painted over, but with a narrow uniform red margin left on the outer edges of the bands. No explanation can be offered.

pyrorhynchus (Fulgora) – Arcana, signature B$_7$, Pl. 4, fig. 1.

= *Pyrops pyrorhynchus* (Donovan, 1800). Author not stated by Perry, who referred to this species as the Bengal Fire Fly.

vitrea (Arcuata) Perry, 1811 (1 January), signature Cc$_7$, Pl. 51.

= *Attacus atlas* (Linnaeus, 1758), the Atlas Moth. Plate legend is *Phalaena.* For a discussion of this synonymy, see under *Arcuata* above. Although stated by Perry to be a "native of South America," this is a southeastern Asia species.

volcanica (Papilio) Perry, 1811 (1 September), signature Rr$_{3-4}$, Pl. 82.

= *Phoebis argante* (Fabricius, 1775). Perry's species name appears on the plate and on both pages of text. The only insect for which both dorsal and ventral views are shown. It is correctly stated by Perry to be a "native of the Rio de la Plata in South America and of Peru." It is notable that Sherborn did not give a signature citation for this species. Mathews & Iredale (1912: 16) listed it as Plate LXXXI. See the discussion herein under Collation.

Trilobita

There are two plates of trilobites in the *Arcana* illustrating three species, all stated to be from "the lime-stone rocks of Dudley in Staffordshire," the well-known Silurian Wenlock Limestone of Dudley. For them Perry introduced a new genus and three new species names. As these names are earlier than any other available names for Dudley trilobites, they could pose problems. Perry's figures are somewhat stylized. They are being studied by trilobite specialists who will take whatever action is necessary to dispose of Perry's unused names. The listing of these nomina herein is not to be construed as a reintroduction but as a listing in the sense of Article 23.9.6 as they are not here associated with any other named genera or species.

Monoculithos Perry, 1810 (1 November), Plate 42, signature Y$_3$.

Type species, by monotypy, *Monoculithos gigantea* Perry, 1810.

gigantea (Monoculithos) Perry, 1810 (1 November), Pl. 42, signature Y_3.

hexamorphus (Monoculithos) Perry, 1810 (1 Dec.), Pl. 48, fig. 2, signature Aa_8.

polymorphus (Monoculithos) Perry, 1810 (1 Dec.), Pl. 48, fig. 1, signature Aa_7.

Mollusca

There are 22 plates of mollusks in the *Arcana*. They are only briefly listed here as they were treated in detail by Petit (2003), which should be consulted for details. Six genus group and three species group molluscan taxa in current usage date from their introduction in the *Arcana*. *Biplex, Hexaplex, Monoplex, Polyplex* and *Proscenula* first appeared in the *Arcana* as genera, with short descriptions but without species, which were then assigned to them in the *Conchology*. One genus group name and one species group name have since been suppressed by action of the ICZN. Since the paper on the mollusks was written the *Arcana* covers have been discovered, necessitating the addition of one name, *Conus dispartus,* that appeared on a cover as a nude name.

Aculea Perry, 1810 (1 April), signature H_5, Pl. 15
Suppressed by I.C.Z.N. Opinion 1677 (1992); = *Haustator* Montfort, 1810.

Aranea Perry, 1810 (1 December), signature Aa_5, Pl. 47.
A junior homonym of *Aranea* Linnaeus, 1758 (Arachnida). A synonym of *Murex* s.s.

Biplex Perry, 1810 (1 June), signature M_7.
Type species, by subsequent designation (Gray 1847: 133), *Biplex perca* Perry, 1811.

Hexaplex Perry, 1810 (1 June), signature M_7.
Type species, by subsequent designation (Iredale 1915: 469), *Hexaplex foliacea* Perry, 1811. = *H. cichoreus* (Gmelin, 1791); = *Murex cichoreum* Gmelin, 1791.

Monoplex Perry, 1810 (1 June), signature M_7.
Type species, by subsequent designation (Dall 1904: 138), *Monoplex australasiae* Perry, 1811. = *Cymatium parthenopeum* (Salis Marschlins, 1793).

Pinus Perry, 1811 (1 September), Pl. 84.
Listed as a genus by Sherborn (1928: 4985) but now treated as a species name. See *Rostellaria pinus* herein.

Polyplex Perry, 1810 (1 June), signature M_7.
Type species, by subsequent designation (I.C.Z.N. 1970, Opinion 911), *Polyplex bulbosa* Perry, 1811. = *Trophon geversianus* (Pallas, 1774); = *Trophon* Montfort, 1810.

Pomacea Perry, 1810 (1 March), signature G_5, Pl. 12.
Type species, by monotypy, *Pomacea maculata* Perry, 1810.

Proscenula Perry, 1810 (1 January), signature B$_5$.
   Three species were described in *Proscenula* by Perry in 1811 in the *Conchology*. All are obviously *Crepidula* but none are identifiable to species. See discussion in Petit 2003: 22.

Septa Perry, 1810 (1 January), signature B$_2$, Pl. *3, fig. 2.*
   Type species, by monotypy, Septa scarlatina Perry, 1810. = Cymatium (Septa) rubecula (Linnaeus, 1758).

Triplex Perry, 1810 (1 June), signature M$_7$, Pl. 23.
   Type species, by monotypy, *Triplex foliatus* Perry, 1810. The species name was declared a *nomen oblitum* in I.C.Z.N. Opinion 911 (1970) but the genus name is in current use as a subgenus of *Chicoreus*.

Volutella Perry, 1810 (1 January), signature B$_2$, Pl. 3, fig. 1.
   Type species, by monotypy, Volutella divergens Perry, 1810. = *Voluta muricata* Born, 1778; = *Vasum* Röding, 1798.

angulata (Aculea) Perry, 1810 (1 April), signature H$_5$, Pl. 15, fig. 2.
   = *Turritella (Haustator) imbricataria* Lamarck, 1804.

angulatus (Conus) Perry, 1810 (1 April), signature H$_5$, Pl. 15, fig. 1.
   = *Conus deperditus* Bruguière, 1792.

apiaria (Trochus) Perry, 1810 (1 January), signature B$_3$, Pl. 3, fig. 4.
   = *Tectus fenestratus* (Gmelin, 1791).

bandatus (Conus) Perry, 1811 (1 August), signature Pp$_7$, Pl. 80.
   Nomen dubium. ? = Conus bandatus Perry, 1811 (*Conchology,* 1 April 1811).

conspicua (Aranea) Perry, 1811 (1 February), Pl. 54.
   = *Bolinus cornutus* (Linnaeus, 1758).

dilatum (Buccinum) Perry, 1811 (1 March), Pl. 58.
   = *Buccinum orbiculare* Perry, 1811, *q.v.*

disjuncta (Scalaria) Perry, 1810 (1 July), signature P$_1$, Pl. 28, center figure, labeled *Scalaria*.
   = *Epitonium scalare* (Linnaeus, 1758).

dispartus (Conus) Perry, 1810 (1 September), back cover of number IX.
   Nomen nudum. Included in list of species to appear in the October issue but was published there as *Conus particolor.*

distentum (Buccinum) Perry, 1811 (1 May), signature B$_3$, Pl. 66.
   Nomen dubium.

divergens (Strombus) Perry, 1811 (1 July), signature Nn$_5$, Pl. 75.
> = *Lambis chiragra arthritica* Röding, 1798. Named *S. divergens* on the plate and *S. nigricans* in the text.

divergens (Volutella) Perry, 1810 (1 January), Pl. 3, fig. 1.
> = *Vasum muricatum* Born, 1778.

flavicunda (Triplex) Perry, 1810 (1 September), signature S$_7$, Pl. 35.
> = *Chicoreus brunneus* (Link, 1807).

foliatus (Triplex) Perry, 1810 (1 June), signature M$_7$, Pl. 23.
> *Nomen oblitum.* See I.C.Z.N. Opinion 911 (1970) and under genus *Triplex* above.

gloriamaris (Conus) Perry, 1810 (1 April), Pl. 16.
> = *Conus textile* Linnaeus, 1758. Not *C. gloriamaris* Chemnitz, 1777.

gracilis (Aranea) Perry, 1810 (1 December), signature Aa$_5$, Pl. 47.
> = *Murex pecten* (Lightfoot, 1786).

laevis (Cerithium) Perry, 1810 (1 April), signature H$_5$, Pl. 15, fig. 3.
> = *Rhinoclavis (Pseudovertagus) striatus* (Bruguière, 1792).

maculata *(Pomacea)* Perry, 1810 (1 March), signature G$_5$, Pl. 12.
> = *Pomacea maculata* Perry, 1810.

nigricans (Strombus) Perry, 1811 (1 July), signature Nn$_5$, Pl. 75.
> = *Lambis chiragra arthritica* Röding, 1798. Named *S. divergens* on the plate and *S. nigricans* in the text.

orbiculare (Buccinum) Perry, 1811 (1 March), signature Gg$_5$, Pl. 58.
> = *Tonna dolium* (Linnaeus, 1758). Named *B. dilatum* on the plate, but the text is headed *B. orbiculare.* The latter name was given precedence by Mathews & Iredale (1912: 13).

pacifica (Voluta) Perry, 1810 (1 August), signature Q$_5$, Pl. 30.
> = *Alcithoe arabica* (Gmelin, 1791).

particolor (Conus) Perry, 1810 (1 October), signature U$_5$, Pl. 39; (*particolar* on plate).
> = *Conus aulicus* Linnaeus, 1758.

phasianus (Bulimus) Perry, 1810 (1 November), signature Y$_7$, Pl. 43.
> = *Phasianella australis* (Gmelin, 1791).

pinus *(Rostellaria)* Perry, 1811 (1 September), signature Rr$_{7-8}$, Pl. 84.
> = *Clavalithes pinus* (Perry, 1811), a senior subjective synonym of *C. macrospira* (Cossmann, 1889).

rubicunda (Rostellaria) Perry, 1810 (1 January), signature B$_3$, Pl. 3, fig. 3.
Nomen dubium.

rubicunda (Triplex) Perry, 1810 (1 September), signature S$_7$, Pl. 35.
= *Chicoreus brunneus* (Link, 1807).

sanguineum (Pecten) Perry, 1811 (1 June), signature Ll$_5$, Pl. 71.
Nomen dubium.

scarlatina (Septa) Perry, 1810 (1 January), signature B$_2$, Pl. 3, fig, 2.
= *Cymatium rubeculum* (Linnaeus, 1758).

solitaris (Strombus) Perry, 1811 (1 January), signature Dd$_1$, Pl. 52.
= *Strombus gallus* Linnaeus, 1758.

tritonia (Septa) Perry, 1810 (1 February), signature D$_5$, Pl. 6.
= *Charonia tritonis* (Linnaeus, 1758).

verrucosa *(Cassis)* Perry, 1810 (1 April), signature H$_5$, Pl. 15, fig. 4.
= *Cassis (Merionella) verrucosa* Perry, 1810. Is *C. cancellata* Lamarck, 1803, not *C. cancellata* Röding, 1798.

zebra (Bulimus) – *Arcana,* 1810, signature L$_5$, Pl. 19.
= *Achatina zebra* (Gmelin, 1791).

zebra (Trochus) Perry, 1811 (1 April), signature Ii$_5$, Pl. 62.
= *Trochus niloticus* Linnaeus, 1758.

Echinodermata

There are two plates of echinoids in the *Arcana,* each depicting a regular sea urchin with spines in place. A third name introduced in the text cannot be identified even to genus.

castaneus (Echinus) Perry, 1810 (1 September), signature S$_5$, Pl. 34.
= *Heterocentrotus mamillatus* (Linnaeus, 1758) *fide* Mortensen (1943: 409, who gave the author as "Parry"), a synonymy confirmed by Andrew Smith. Mortensen (1943: 419) stated that *mamillatus* is historically correct as it was so spelled by Linnaeus, but he used *mammillatus,* stating that to be the "linguistically correct form, since there can be no doubt it is derived from the Latin word *mammilla.*" He was not the first to use that spelling and his chresonymy shows both spellings have been used almost equally. As to the derivation of the species name, Linnaeus clearly took it from Klein, listing Klein's *"Cidaris mamillata"* under this and one other species. Gmelin (1791: 3175) spelled it *mammillatus* and cited a later edition of Klein. The two different Klein editions have not been checked, as the name as originally spelled is not subject to

emendation. Even had Linnaeus changed the spelling in the 1767 12th Edition of his *Systema,* such an emendation would not be valid under the present Code.

The spelling *mammillatus* was adopted by Rowe & Gates (1995), who list Gmelin as the source of the emended spelling and cite Mortensen. A search of the web (12 January 2006) shows that the spelling *mammillatus* is currently in much greater usage than *mamillatus* and it is possible that this emended spelling can be retained under Article 33.3.1. This determination is left to echinoid specialists.

sceptriferus (Echinus) Perry, 1810 (1 September), signature S_6.

= *Nomen dubium.* Mentioned under *Echinus castaneus* by Perry. He indicated that it had been in the collection of the Duchess of Portland and "at present belongs to . . . Mr. Jennings, of Chelsea, and which we propose to delineate if possible in a future number, it is remarkable for having jointed spines, and is allowed to be exceedingly rare . . ." A search of the Portland Catalogue ([Lightfoot] 1786) failed to locate any echinoid that could be identified as this species. Were it not for the statement about jointed spines this could be dismissed as a *nomen nudum* but the Code does not qualify or quantify "description." Perry's description is odd as he must have known that echinoid spines have a joint at the base and are mobile.

stellaris (Echinus) Perry, 1810 (1 October), signature U_3, Pl. 38.

= *Nomen dubium.* This is an unidentifiable species, probably in the genus *Phyllacanthus fide* Andrew Smith.

Pisces

Only eight plates feature fish, but one genus group name and one species group name introduced by Perry are in current usage.

Congiopodus Perry, 1811 (1 February), signature Ee_8, Pl. 55.

Type species, by monotypy, *Congiopodus percatus* Perry, 1811. = *C. torvus* (Gronow, 1772) *fide* Dr. Paolo Parenti. It was incorrectly listed as being on signature Dd_8 by Sherborn (1925: 1464) and Neave (1939: 816).

Hippocampus Perry, 1810 (1 May), signature K_5, Pl. 18.

Type species, by monotypy, *Hippocampus foliatus* (Shaw, 1804). = *Phyllopteryx taeniolatus* (Lacepède, 1804). This genus name is preoccupied by *Hippocampus* Rafinesque, which evidently has one month's priority.

Dolphin – *Arcana,* 1810, signature H_{1-2}, Pl. 13.

= *Coryphaena hippurus* Linnaeus, 1758. The plate is a good illustration of the fish, but Perry's description and discussion contain a mixture of the features of the fish and the mammal (Porpoise) that is sometimes referred to as a Dolphin. Perry was aware of the difference as in his opening sentence he states that "the Dolphin bears a considerable resemblance in its external form to the Porpoise" but then continues with a mixture of characters. He correctly states that the Dolphin is more slender and his description of its color, especially the change in coloration upon death, accurately describes the fish. He attributes a size of "eight or ten feet in length" to the Dolphin but its maximum size is about two meters. The figure is stated to have been drawn from a specimen "about three feet in length" in Mr. Bullock's Museum. Perry's second paragraph accurately describes the head, fins and color markings of the fish. While the Dolphin does prey on other fish, as does the Porpoise, it does not attack whales as alleged.

The latter part of Perry's discussion is almost entirely about the Porpoise, as he relates its attachment to mankind, its close following of vessels and Arion being carried ashore on its back.

No binominal name was given by Perry but there are two pre-Linnaean names in the text heading: *Delphinus* of Pliny and *Porcus Marinus* of Sibbald, both of which refer to the mammal. It was listed under "Mammalia" by Mathews & Iredale (1912: 14) without comment. This is a species of the open seas.

bandatus (Sparus) Perry, 1810 (1 February), signature E_3, Pl. 8.
= *Cheilinus fasciatus* (Bloch, 1791) *fide* Parenti & Randall, 2000: 8. This species ranges through the Indo-West Pacific.

depressus (Stromateus) Perry, 1810 (1 July), signature O_5, Pl. 26.
Nomen oblitum. = *Selene setapinnis* (Mitchell, 1815). Perry's species is considered to be an unused senior synonym of *Selene setapinnis* (Mitchell, 1815), a well-known and frequently cited Western Atlantic species. Perry's name is listed in the synonymy of *S. setapinnis* (Mitchell, 1815) by Eschmeyer (2007) as a *nomen oblitum.*

erectus *(Hippocampus)* Perry, 1810 (1 December), signature Aa_1, Pl. 45.
= *Hippocampus erectus* Perry, 1810. A Western Atlantic species, stated by Perry to be a "native of the American seas." Listed as a valid species by Eschmeyer (1998: 537).

faber (Zeus) – *Arcana*, 1811, signature Pp_{5-6}, Pl. 79.
= *Zeus faber* Linnaeus, 1758. The John Dory. No author was cited by Perry for this well- known species that has a cosmopolitan distribution.

foliatus (Hippocampus) – *Arcana*, 1810, signature K_5, Pl. 18.
= *Hippocampus foliatus* (Shaw, 1804). = *Phyllopteryx taeniolatus* (Lacepède, 1804) *fide* Eschmeyer (2003). Author not cited by Perry. This is an Australian species as stated by Perry. Type species of *Hippocampus* Perry, *q .v.*

niloticus (Esox) Perry, 1811 (1 April; plate dated March), signature Kk_1, Pl. 64.
= *Lepisosteus osseus* (Linnaeus, 1758) *fide* Dr. Paolo Parenti. Perry's locality of "River Nile" is incorrect, as this is a North American freshwater species.

percatus (Congiopodus) Perry, 1811 (1 February), signature Ee$_8$ [Ee$_7$ in reissue], Pl. 55.

    = *C. torvus* (Gronow, 1772) *fide* Dr. Paolo Parenti. Incorrectly listed as on signature Dd$_8$ by Sherborn (1928: 4845). This species is from southern Africa. The type species of *Congiopodus* Perry, *q.v.*

Reptilia

There are only three species of reptiles in the *Arcana*. They were treated by Bauer & Petit (2004), who reproduced the plates.

horridus (Crotalus) – *Arcana,* 1810 [1 February], signature D$_{1-4}$, Pl. 5.

    = *Crotalus horridus* Linnaeus, 1758. This is one of the few species for which authorship is attributed by Perry. Perry's plate was reproduced and the problematic nature of its identification discussed by Bauer & Petit (2004: 10, fig. 2).

pallida (Chamaeleo) Perry, 1810 [1 July], signature O$_1$, Pl. 25.

    ? = *Chamaeleo senegalensis* Daudin, 1802. The identification of Perry's figure and this probable synonymy were discussed in detail by Bauer & Petit (2004: 11, fig. 3), who also reproduced the plate.

panama (Testudo) Perry, 1810 [1 September], signature S$_1$, Pl. 33.

    Nomen oblitum. Considered to be a senior subjective synonym of *Trachemys venusta* (Gray, 1856), it was shown to be a *nomen oblitum* by Bauer & Petit (2004: 14, fig. 4), who reproduced the plate.

Aves

There are fourteen plates of birds in the *Arcana*. As Mathews and Iredale were especially interested in birds and mollusks it is surprising that one bird plate (Plate 44) was omitted in their compilation. One bird, the "Condor Vulture," was not identified by a Latin name.

Cassowara Perry, 1811 (1 September), signature Rr$_5$.

    Type species, by monotypy, *Cassowara eximia* Perry, 1811. A junior subjective synonym of *Casuarius* Brisson 1760.

Condor Vulture – *Arcana,* 1810, signatures D$_{7-8}$, E$_{1-2}$, Pl. 7.

    = *Vultur gryphus* Linnaeus, 1758. The Andean Condor. Probably the most striking plate in the *Arcana,* the vulture is shown carrying a rather large child, which is impossible as American Vultures cannot grip with their feet or carry anything.

africanus (Pelicanus) – *Arcana,* 1811, signature Ff$_2$ [Ff$_1$ in reissue], Pl. 56.

    ? = *Pelecanus onocrotalus* Linnaeus, 1758, the Great White Pelican. Name not attributed by Perry but certainly intended to be *Pelecanus africanus* Gmelin, 1789.

Mathews & Iredale (1912: 9) stated that it would enter into the synonymy of *P. rufescens* Gmelin, but it is considered by Michael Walters to more probably be *P. onocrotalus* Linnaeus. Perry did not introduce the spelling *Pelicanus* as it is an unjustified emendation of *Pelecanus* Linnaeus by Boddaert 1783 *fide* Neave (1940: III: 640).

aurantius (Lanius) Perry, 1811 (1 June), signature Ll$_3$, Pl. 70.
Nomen dubium. Stated to be a native of "Buenos Ayres" by Perry. Mathews & Iredale (1912: 9) considered it to be an indeterminable species of *Thamnophilus,* an assessment with which Michael Walters concurs. There is no legend on the plate.

coerulea (Cerithia) Perry, 1810 (1 November), signature Z$_1$, Pl. 44.
? = *Cyanerpes cyaneus* (Linnaeus, 1766). *Cerithia* was listed by both Sherborn and Neave as an error for *Certhia* Linnaeus with a query. It is here considered to be a misspelling and not a new taxon. The species name *coerulea* is spelled with a ligature that appears to be "oe" on the plate and text heading and Sherborn lists it as *C. coerulea.* However, within the text it is clearly "ae" in its only appearance there (signature Z$_2$). The difference in, or more correctly, indifference to, ligatures in the *Arcana* was addressed by Petit (2003: 19) and *coerulea* is here accepted as correct. Referred to by Perry simply as a Creeper, it is thought to be the Red-legged Honeycreeper, *Cyanerpes cyaneus* (Linnaeus, 1766). This species was overlooked by Mathews & Iredale, 1912.

coronata (Ardea) Perry, 1811 (1 August), signature Pp$_3$, Pl. 78.
= *Balearica pavonina* (Linnaeus, 1758). This synonymy was shown by Mathews & Iredale (1912: 9). Known as the Crowned Crane, it was referred to by that name in Perry's text but as Crown Crane on the plate. A native of Africa.

eximia (Cassowara) Perry, 1811 (1 September), signature Rr$_{5-6}$, Pl. 83.
= *Casuarius casuarius* Linnaeus, 1758. Supposedly from South America but considered by Mathews & Iredale (1912: 9) to be a New Guinea form not previously recognized. It now considered a junior subjective synonym of *C. casuarius* Linnaeus, 1758, the Southern Cassowary, *fide* Michael Walters.

militaris (Ara) – *Arcana,* 1810, signatures L$_{7-8}$, Pl. 20.
= *Ara militaris* (Linnaeus, 1766). Name not attributed by Perry, but this is Linnaeus' species, the Military Macaw. This species has a wide range in South America.

niger (Anas) – *Arcana,* 1811, signature Gg$_{7-8}$, Pl. 59.
= *Cygnus atratus* (Latham, 1790). The name was not attributed by Perry but his text indicates that it was intended for *Anas cygnus nigra* Linnaeus, 1758. Only the name Black Swan appears on the plate.

nonpareil (Psittacus) Perry, 1810 (1 March), signature F$_1$, Pl. 9.

= *Platycercus eximius* (Shaw, 1792). This synonymy was noted by Mathews & Iredale (1912: 8). Perry called this the Nonpareil Parrott (misspelled as Nonpariel on the plate) but it is now known as the Eastern Rosella. In the discussion under this species Perry mentioned that Linnaeus divided the birds into six orders which he then listed, though omitting order number 3. This is an Australian species that was introduced to New Zealand in the early 20th Century.

papuensis (Psittacus) – Arcana, 1810, signatures T$_{1-2}$, Pl. 36.

= *Charmosyna papou* (Scopoli, 1786). The name was not attributed by Perry but it is certainly intended to be *Psittacus papuensis* Gmelin, 1788, a junior synonym of *C. papou* (Scopoli). Referred to by Perry as the Papuan Lory, the vernacular name in current usage is Papuan Lorikeet. As implied by the name, this species is a native of New Guinea and surrounding islands.

regia (Paradisea) – Arcana, 1810, signature Q$_1$, Pl. 29.

= *Cicinnurus regius* (Linnaeus, 1758). The name was not attributed by Perry, who referred to it as the King Bird of Paradise (misspelled Paradice on the plate), a vernacular name still in use for this New Guinea species. Perry's figure is of a "singularly fine specimen in Mr. Bullock's Museum." When Bullock's Museum was auctioned, on the twenty-second day's sale (4 June 1819) what was probably this specimen, described as "the largest specimen of the bird known, and in the highest preservation," was sold for 4 pounds 18 shillings.

rubicunda *(Ardea)* Perry, 1810 (1 June), signature M$_5$, Pl. 22.

= *Grus rubicunda* (Perry, 1810). Now know as the Brolga, the plate legend is New Holland Crane but in the text it is called the Red-Headed Crane of New Holland. This is the only bird name introduced by Perry that is now in use. The figure was stated by Perry to be drawn from a specimen in Bullock's Museum. That specimen is believed to be one sold as Lot 12 on the fifteenth day of the Bullock Museum auction (25 May 1819) to Dr. Walter Adam, who was buying on behalf of the Edinburgh Museum (now National Museums of Scotland). Unfortunately no bird specimens purchased at the Bullock sale are still extant in the Edinburgh collections (Sweet, 1970: 32).

rubra (Tringa) Perry, 1811 (1 January), signature Cc$_5$, Pl. 50.

? = *Phalaropus lobatus* (Linnaeus, 1758), the Red-necked Phalarope. The name on the plate is simply Red Phalarope. Perry showed his name to be a variety and it must be assumed that he considered it to be a variety of *Tringa hyperborea* Linnaeus, 1766 as the only species mentioned in the text is a description of that species from the "work of Mr. Bewick." *Phalaropus lobatus* (Linnaeus, 1758) is an earlier synonym.

The figure and description do not allow for definitive identification and this vernacular name has been used for both the Grey (= *P. fulicarius* (Linnaeus, 1758)) and the Red-necked Phalarope. This is a European species.

viridis (Psittacus) Perry, 1810 (1 March), signature G$_1$, Pl. 11.
= *Pezoporus wallicus* (Kerr, 1792). Although there is a *Psittacus viridis* Latham, 1790, Perry intended this to be a new name as he stated that "the present bird is the Green Spotted Parroquet of New Holland, and is supposed to be hitherto undescribed." Mathews & Iredale (1912: 8) considered it to be a synonym of *Pezoporus terrestris* (Shaw, 1793), a synonym of the earlier *P. wallicus* (Kerr, 1792). This Australian endemic is now known as the Ground Parrot.

Mammalia

There are 22 plates of mammals in the *Arcana,* the same number as are devoted to Mollusca. Perry introduced four new genus group names and twelve new species group names for mammals, but only one of each is in current usage. Two mammals were illustrated without binominal names, one of which is the first published figure of a Koala.

Antelopa —Arcana, 1811, signature Ee$_1$, Pl. 53.
= *Antilope* Pallas, 1766. Mathews & Iredale (1912: 14) stated that *Antelopa* was newly introduced by Perry. However, it is listed by Neave (1939: 222) as "pro *Antilope* Pallas" and is here considered to be a misspelling as Perry referred to Pallas' work in his discussion. See below under the species *montana.*

Babyrousa Perry, 1811 (1 May), signature C$_1$, Pl. 67.
Type species, by monotypy, *Babyrousa quadricornua* Perry, 1811 (= *Sus babyrussa* Linnaeus, 1758). Mathews & Iredale (1912: 14) showed this to be the earliest name for the genus. The name is spelled *Babyroussa* on Plate 67 in error.

Guanaco Perry, 1811 (1 May), signature C$_3$, Pl. 68.
Type species, by monotypy, *Guanaco patagonia* Perry, 1811. A junior subjective synonym of *Lama* Cuvier, 1800. As shown below under *G. patagonia,* Perry's figure is not of a Guanaco but of a Llama, *Lama glama* (Linnaeus, 1758). This is a case where the type species could be selected as the species he actually had instead of the species he thought he had, but it makes no difference as they are now considered to be congeneric. *Guanaco* was thought by Mathews & Iredale (1912: 14–15) to be the oldest name for the genus. Neither Perry's genus nor species are listed in Wilson & Reeder, 1993.

Opossum Perry, 1810 (1 June), signature M$_1$, Pl. 21.
Type species, by monotypy, *Opossum hirsutum* Perry, 1810. = *Vombatus ursinus* (Shaw, 1800). A junior subjective synonym of *Vombatus* Geoffroy, 1803. Mathews & Iredale

(1912: 14) stated that this was the only instance noted of *Opossum* being used as a generic name, but it was used in the August issue by Perry for a quite different animal. They also noted that Perry's name was junior to *Phascolomis* Geoffroy, 1803, a name itself now considered junior to *Vombatus* Geoffroy 1803 (*fide* Dawson in Walton, 1988: 49). Groves (in Wilson & Reeder, 1993: 45) lists *Oppossum* [*sic*] in the synonymy of *Vombatus* and although not attributed it is probably a misspelling of Perry's *Opossum* as the word *hirsutus,* not attributed to either genus or author, appears among the species-level synonyms. As Perry is not in the References Cited, the author's intention is unclear.

Koalo – *Arcana,* 1810, signature K_{1-4}, Pl. 17.

= *Phascolarctos cinereus* (Goldfuss, 1817). The plate legend is simply Koalo, and text caption is Koalo, or New Holland Sloth. Perry's arrangements of names and terminology were very peculiar and there is no clear introduction of a new name here. Possibly Perry intended to introduce it as a species of *Bradypus* but did not actually do so. It is unfortunate that Perry missed the opportunity of naming one of the world's most popular animals.

Perry's Koalo plate is featured on the web site of the State Library of South Australia, where it is stated to be the first known published illustration of a Koala, though it is dated there as 1811 instead of 1810.

Spotted Hyena – *Arcana,* 1811, signature Rr_{1-2}, Pl. 81.

= *Crocuta crocuta* (Erxleben, 1777). Legend on plate is Spotted Hyena. In the text heading and within the text the name is spelled Hyaena with the "ae" as a ligature. Either spelling is correct, Hyena being more commonly used in the United States. A native of Africa, where it is widely distributed, this is the only Hyena with spots.

aries (Ovis) – *Arcana,* 1811, signature Pp_{1-2}, Pl. 77.

= *Ovis aries* Linnaeus, 1758. Legend on plate is Ægyptian Ram. This species is domesticated, with worldwide distribution.

camelopardalis (Camelus) – *Arcana,* 1811, signature Cc_{1-4}, Pl. 49.

= *Giraffa camelopardalis* (Linnaeus, 1758). Not attributed by Perry, but obviously Linnaeus' species. The figure was drawn from a seventeen-foot specimen in Bullock's Museum. The Museum was sold at auction in 1819 and this specimen, which can be identified by size as well as by having been stated by both Perry and Bullock to be the "finest specimen in Europe," was purchased by Dr. Adam for the Edinburgh Museum for the then very high sum of £24. It is no longer extant (Sweet, 1970: 32, confirmed by Dr A. Kichener).

fossilis (Cervus) Perry, 1811 (1 June), signature Ll_7, Pl. 72.

= *Megaloceros giganteus* (Blumenbach, 1799). Perry's figure, captioned "Fossil Elk" was drawn from a specimen of the "Irish Elk" said to be "twelve feet in the girth."

Gould (1977), in an interesting essay on "The Misnamed, Mistreated, and Mis-understood Irish Elk", stated that Cuvier proved in 1812 that it was not like any modern animal. Perry had already discussed the "two Elks now known to exist" and then stated of the 'Fossil Elk' that "we must consider it as an antidiluvian, and that it has become extinct, (perhaps at that period) is evident, since no animal has so great a width of horns as the above, in any known part of the globe . . ." Of course, others had also stated that it was "fossil" but Cuvier was the first to provide scientific evidence that it was unlike any living species.

Perry recounted the story of an Irish gentleman who, upon being asked for an explanation of this phenomenon, very gravely answered: "that he thought it not a natural but an accidental cause, for that his horns were much too wide and spreading to admit of passing into the door of Noah's Ark."

gigas (Elephas) Perry, 1811 (1 April), signature Ii$_1$, Pl. 61.

  = *Elephas maximus* Linnaeus, 1758 *fide* A. Shoemaker. Wilson (in Wilson & Reeder, 1993: 367) lists as synonyms of *E. maximus* a string of species group names, among which is *gigas*. Those names are not identified by either genus or author and it is thus impossible to determine if Perry's *E. gigas* is intended, especially as Perry is not listed in the Literature Cited.

  This figure was drawn from an elephant "in the menagerie of Mr. Polito in the Strand." This later became Polito's Royal Menagerie, located about where the Strand Palace Hotel now stands. The elephant was named Chunee and was well-known, as he was taken out of his cramped quarters each night and walked along the Strand. In 1826 Chunee went berserk, suffering from a diseased tusk. After three days of massive cage battering, a hastily assembled file of soldiers was brought in and they fired no fewer than 252 musket balls into his body before running out of ammunition. Chunee then succumbed to a harpoon thrust. His skeleton, skull riddled with bullet scars, was a feature at the museum of the College of Surgeons for many years until it was destroyed in 1941 in a bombing raid (Brightwell, 1952; cited from Hancocks, 2001: 37).

hirsutum (Opossum) Perry, 1810 (1 June), signature M$_1$, Pl. 21.

  = *Vombatus ursinus* (Shaw, 1800) *fide* Dawson (in Walton, 1988: 49–50). The plate legend is simply Wombach. Perry's species name was stated by Mathews & Iredale (1912: 14) to be the earliest for what they termed the New South Wales Wombat, but it is now known that Shaw's name is earlier. See discussion under *Opossum* above.

jacchus (Sapajus) – Arcana, 1810, signatures U$_{7-8}$, X$_{1-2}$, Pl. 40.

  = *Callithrix jacchus* (Linnaeus, 1758). Author not stated by Perry. Legend on plate is Sanglin Monkey. This is a Brazilian species.

leo (Felis) – *Arcana,* 1811, signature Gg$_{1-4}$, Pl. 57.

    = *Panthera leo* (Linnaeus, 1758). Legend on the plate is simply Lion.

leopardis (Felis) – *Arcana,* 1811, signature B$_{1-2}$, Pl. 65.

    = *Panthera pardus* (Linnaeus, 1758). Name not attributed but certainly a misspelling of *Felis leopardus* Schreber, 1777, which was shown to be a synonym of Linnaeus' species by Wozencraft (in Wilson & Reeder, 1993: 298). The figure shows some of the dorsal rosettes to have spots within their center, a feature mentioned in the text. However, while this is the normal marking for New World jaguars, only the leopard of Kasmir routinely shows this trait. Perry did not give any source for the animal figured and discussed it in only vague terms other than the description of the markings. His discussion begins with the statement that "the Leopard is very nearly approximated to the Panther, which we have described in a former part of the Arcana, so as to make it difficult to distinguish the two creatures." Also, that "this animal is smaller than the Panther" after which he described the spots. Although spots are shown in center of some of the rosettes on the Panther figure, only the color and shape of the spots is mentioned in the text. See *Felis pantherus* below.

longirostra (Platypus) Perry, 1810 (1 March), signature F$_{6-7}$ (not figured).

    = *Tachyglossus aculeatus* (Shaw, 1792) *fide* Mahoney (in Walton, 1988: 5; as *P. longirostratus* and as *P. longirostrata*). Mathews & Iredale (1912: 13) considered this an earlier name for *Echidna setosa* Cuvier [*sic;* = Geoffroy, 1803], another synonym. Groves (in Wilson & Reeder, 1993: 13) included in the synonymy of *T. aculeatus* (Shaw, 1792), the word *"longirostris"* but it is not attached to a genus or to an author so it cannot be determined if it is an error for Perry's name. In either event *P. longirostra* Perry is a junior synonym. This species is the Short-beaked Echidna.

montana (Antelopa) Perry, 1811 (1 February), signature Ee$_1$, Pl. 53.

    = *Alcelaphus buselaphus* (Pallas, 1766). *Antelopa* is considered to be a *lapsus* for *Antilope* Pallas. The plate legend is Mountain Cow, as is the text caption. Within the text the animal is referred to as the Mountain Antelope. This is the North African nominate form of this species, and is now extinct; subspecific forms are still found in other parts of Africa. Perry's species name is not listed in Wilson & Reeder, 1993.

muscola (Dipus) Perry, 1810 (1 July), signature O$_7$, Pl. 27.

    = *Potorous tridactylus* (Kerr, 1792) *fide* Calaby & Richardson (in Walton, 1988: 58). There is no legend on the plate and Mathews & Iredale were unable to identify the species, now commonly known as the Long-nosed Potoroo. Calaby & Richardson *(op. cit.)* incorrectly state that Perry's figure was based on the description in one of two earlier accounts. Perry referred to one of those accounts and definitely stated that the species figured was different, being the "only . . . specimen . . . known to be in England, and which is in Mr. Bullock's Museum."

opossum (Opossum) Perry, 1810 (1 August), signature R$_1$, Pl. 32.

    = *Acrobates pygmaeus* (Shaw, 1794) *fide* S. Ingleby. The plate legend is Opossum Mouse and the text caption is Opossum Flying Mouse. There is no reference to *Opossum opossum* in either McKay (in Walton, 1988: 98–102) or in Wilson & Reeder (1993). This species is now known as the Feathertail Glider.

pacos (Camelus) – Arcana, 1810, signature H$_{3-4}$, Pl. 14.

    = *Lama pacos* Linnaeus, 1758. Name attributed to Linnaeus by Perry. The plate legend is simply Vicuna.

pantherus (Felis) – Arcana, 1811, signature Ii$_{7-8}$, Pl. 63.

    = *Panthera pardus* (Linnaeus, 1758). Although Perry did not attribute the name he used, it is probable that it was intended for *Felis pantherus* of Erxleben (1777) or of Schreber (1777) [neither seen; both names from Sherborn (1902: 714); neither in Wilson & Reeder (1993)]. The name on the plate is simply Panther. See remarks above under *leopardis.* These two big cats cannot be differentiated from Perry's figures.

paradoxus (Ornithorinxus) – Arcana, 1810, signature F$_5$, Pl. 10.

    = *Ornithorhynchus anatinus* (Shaw, 1799); = *O. paradoxus* Blumenbach, 1800 *fide* Mahoney (in Walton, 1988: 8). *Ornithorinxus* is a misspelling. Authorship was not cited by Perry. The plate legend is simply Platypus. The text caption is "Platypus; or Ornithorinxus paradoxus" without any italics. It is often difficult to determine Perry's intentions but he may have intended to place it in the genus *Platypus* as shown by his later description of *Platypus longirostra* in the discussion. In either event, the species name is a junior synonym.

    The Australasian mammals were, by Perry, "considered as the strongest natural proof to a reasoning mind, that the Flood or Diluvian Overflux of the Ocean, was not universal, for if so, it would be impossible to account for the restoration of each individual species to each particular climate."

patagonia (Guanaco) Perry, 1811 (1 May), signature C$_3$, Pl. 68.

    = *Lama glama* (Linnaeus, 1758) *fide* Alan Shoemaker. The only name on the plate is Guanaco. The animal figured by Perry is not a Guanaco but a Llama—it is one of the many vagaries of systematic nomenclature that the scientific name is spelled with one "l" and the vernacular name with two. The text is headed, in Perry's slightly ambiguous way and without any italics as usual, "Genus – Guanaco. Species – Guanaco, Patagonia." By comparing other Perry headings it is obvious that patagonia is a species name, as listed by Mathews & Iredale (1912: 14), and not a location. Perry often cited binomina in that manner. Neither Perry's genus nor species names appear in Wilson & Reeder (1993).

quadricornua (Babyrousa) Perry, 1811 (1 May), signature C$_1$, Pl. 67.

= *Babyrousa babyrussa* (Linnaeus, 1758) *fide* Alan Shoemaker. Plate legend is *Baby-roussa.* Mathews & Iredale (1912: 14) stated that the figure is of a skull from Amboyna, but Perry stated of the animal that it is "very seldom seen alive in England" and that "it resides near the countries of Amboyna." This is Ambon, one of the Moluccas Islands in Indonesia.

socotrus (Elephas) Perry, 1811 (1 April) signature Iii.

? = *Elephas maximus* Linnaeus, 1758. This species is not figured by Perry but was named under the discussion of *E. gigas* as "a second and different species, which is said to reside in the kingdom of Thibet [*sic*], and being much smaller and of an opposite form, is to be considered as a separate animal from the above, under the title or Name of the Elephas Socotrus." Although listed by Sherborn (1929) and by Mathews & Iredale (1912) the name does not appear in Wilson & Reeder (1993). This is obviously a junior synonym of *E. maximus* Linnaeus.

striatus *(Bradypus)* Perry, 1810 (1 November), signature Y$_1$, Pl. 41.

= *Ictonyx striatus* (Perry, 1810). There is no legend on the plate. Perry stated that the animal was "reported to have been found in South America" but the locality is now known to be sub-Saharan Africa. Perry's taxon is on the Official List of Specific Names in Zoology (I.C.Z.N. 1967, Opinion 818). This is the animal now known as the Zorilla.

tigris (Felis) – *Arcana,* 1811, signature A$_{1-4}$, Pl. 1.

= *Panthera tigris* (Linnaeus, 1758). Name attributed to Linnaeus by Perry. This plate, one of the very few showing an animal in motion, depicts a tiger leaping from the entrance of a cave and over a skeleton.

tridactylus (Dipus) Perry, 1811 (July 1), signature [Nn]$_1$, Pl. 73.

= *Macropus giganteus* Shaw, 1790 *fide* Calaby & Richardson (in Walton, 1988: 67). The spelling in the plate legend is Kanguroo, but in the text caption and in the text it is Kangaroo. This is the Eastern Grey Kangaroo.

zebra (Equus) – *Arcana,* 1811, signature [Nn]$_7$, Pl. 76.

= *Equus zebra* Linnaeus, 1758. The animal figured is Hartmann's Mountain Zebra from Namibia, *Equus zebra hartmannae* Matschie, 1898.

Botany

In the Introduction to the *Arcana* Perry announced his intention to include botanical subjects but the only one included was in the first issue.

Palm Tree – *Arcana,* signature A$_{5-8}$, Pl. 2.

   = *Ceroxylon quindiuense,* the Quindio Wax Palm. Perry's plate, inscribed "Botany Pl. 1" (all four plates in the first issue are "Pl. 1"), bears the legend "Palm Tree." The text heading is "The Ceroxylon, or Palm Tree." Perry stated it to be a native of Quindiu [Peru] which permits identifying it to species although the swelling in the middle of the trunk does not match available photographs. After discussing the wax production of this palm Perry mentioned that it may be "regarded as the Monarch of all the Forests of the World, if its gigantic size can entitle it to that distinction." This is indeed one of the world's tallest palms, reaching a height of 60 meters.

ACKNOWLEDGMENTS

This book would not have been possible without the assistance and cooperation of a number of people. The following specialists advised on the various groups of animals, providing identifications, synonymies, references to synonymies and supporting literature:

Insecta: As insect taxonomy is very specialized, a number of persons assisted in bringing Perry's taxa up to date. Dr. Brian R. Pitkin not only contributed comments and suggestions, but also coordinated the efforts of Dr. Ian Kitching, Mrs. Judith A. Marshall, Mr. Mick Webb and Mr. Phil Ackery, all of The Natural History Museum, London. Dr. Gerardo Lamas, Museo de Historia Natural, Universidad Nacional Mayor de San Marcos, Lima, Peru also contributed and was especially helpful in supplying references.

Trilobita: Dr. Philip D. Lane, Keele University and Dr. Derek J. Siveter, University Museum of Natural History, Oxford corresponded and discussed the trilobites and furnished literature. Dr. Richard A. Fortey, Natural History Museum, London, was an early contact for this group and also furnished valuable information. These specialists are studying Perry's trilobite taxa and will take any action deemed necessary to conserve names in long-standing usage.

Echinoids: Andrew Smith, The Natural History Museum, London and Cynthia Ahearn, National Museum of Natural History, Smithsonian Institution, Washington, D.C. provided identifications and references.

Pisces: Dr. William N. Eschmeyer, California Academy of Sciences, San Francisco, California provided identifications and literature as did Dr. William F. Smith-Vaniz, U. S. Geological Survey, Gainesville, Florida. Dr. Richard L. Pyle, Bishop Museum, Honolulu and Dr. Paolo Parenti, Università di Milano-Bicocca also rendered valuable advice.

Reptilia: Dr. Aaron M. Bauer, Villanova University, Villanova, identified the *Arcana* reptiles and was first author on a recent paper treating them (Bauer & Petit, 2004).

Aves: The assistance of Michael Walters, The Natural History Museum at Walter Rothschild Zoological Museum, Tring, who provided the identifications, synonymies, and other helpful information is gratefully acknowledged. Dr. Frank Steinheimer, Museum für Naturkunde der Humboldt-Universität zu Berlin provided information on birds from Bullock's Museum.

Mammals: Dr. Sandy Ingleby, Australian Museum, Sydney, provided identifications of the Australasian species and literature references. The other mammals were identified by Mr. Alan Shoemaker, Red List Authority, IUCN Tapir Specialist Group, Columbia, SC. Dr. Andrew Kitchner, National Museums of Scotland, Edinburgh, kindly corresponded about a Bullock specimen.

Botany: Dr. N. Dwight Camper, Clemson University, Clemson, South Carolina, identified the tree illustrated and discussed by Perry.

Many librarians offered information on copies of the *Arcana* in their care, often going to much trouble to look up specific items and to provide collations. Their contributions were invaluable. For want of a perfect system they are here acknowledged alphabetically: Vi Adlam, Science & Technology Librarian, State Library of Western Australia, Perth; Stella Brecknell, Oxford Museum Library, Oxford; Carol Cantrel, Research Library, Australian Museum, Sydney; Patrick Condron, Brownless Biomedical Library, University of Melbourne; David W. Corson, Cornell University; Patricia Darley, Research Officer, State Library of South Australia, Adelaide; Dr. C. Daniel Elliott, former Library Director, Academy of Natural Sciences of Philadelphia; Brian Gerrard, Monash University Library, Melbourne; Paul Livingston, Senior Reference Librarian, National Library of Australia, Canberra; Eleanor MacLean, Blacker-Wood Librarian, McGill University, Montreal; Carol Murray, CSIRO Black Mountain Library, Canberra; Claus-Ove Strandberg, Head of the Rare Book Collections, Stockholm University Library; Margaret Triffitt, Western Australia Museum Library, Perth; Alan Ventress, Mitchell Librarian, State Library of New South Wales, Sydney; Priscilla Watson, American Museum of Natural History Library; Anne Xu, Document Supply Services, National Library of Australia, Canberra.

Non-librarians who assisted in obtaining collations and other data are: Dr. Robert H. Cowie, University of Hawaii; Dr. Paula M. Mikkelsen, Paleontological Research Institution, Ithaca, NY; the late Walter Sage, American Museum of Natural History; Michael Walter, The Natural History Museum, Zoological Museum, Tring and Kathie Way, The Natural History Museum, London. Leslie Jessop, Sunderland, UK, recently discovered and collated the copy at Newcastle-upon-Tyne.

Copies of literature were made available by Lorraine Portch and Eliza Walsh, The Natural History Museum, London.

Several antiquarian book dealers furnished information about copies of the *Arcana* handled by their firms and, in cases where copies had been sold to individuals, contacted those individuals requesting that they contact this compiler. They are: Derek McDonnell, Hordern House, Sydney; Julian Wilson, Maggs Bros., Ltd., London; Tony Swann and the late C. K. Swann, Wheldon & Wesley Ltd., last at Leighton Buzzard, UK, but unfortunately no longer in business. Sir Tom Ramsey, Australia and John Robinson, Coulsdon, Surrey, UK, corresponded about copies of the *Arcana* that they once owned.

Mr Scott Jordan has been interested in the *Arcana* for many years and actively searched for copies, contacted and corresponded with many of the libraries holding copies, and made collations of several copies. He also made his copies available for this compilation. His cooperation and assistance has been invaluable.

Special thanks are due to Prof. Hugh Torrens, University of Keele, who kept this writer informed of his progress in developing biographical data on Perry and also furnished much other valuable information and literature. Prof. Torrens examined the Oxford *Arcana* covers and made important observations about the manner in which the *Arcana* was originally issued.

Numerous persons contributed helpful information or advice including: Dr. Robert Burn, Victoria; Dr. Robert H. Cowie, University of Hawaii, Honolulu; S. Peter Dance, Carlisle; Gina Douglas, Linnean Society, London; Nicholas Haberman, Archives, The Natural History Museum, London; Dr. M. G. Harasewych, National Museum of Natural History, Washington; Dr. Alan Kabat, Washington; William G. Lyons, St. Petersburg; Dr. E. Charles Nelson, Cambridgeshire; Dr. Alan P. Peterson, Walla Walla; Dr. Gary Rosenberg, Academy of Natural Sciences of Philadelphia; Dr. Martin A. Snyder, Villanova; Kathie Way, The Natural History Museum, London. Susan Birch, Image Technician, Oxford University Museum of Natural History, made digital copies of the *Arcana* covers.

The author gratefully acknowledges the assistance of all those named here, and assumes sole responsibility for any errors in the present work.

LITERATURE CITED

Bauer, A. M. & Petit, R. E. 2004. On the herpetology of Perry's *Arcana* and two forgotten reptile names. Newsletter and Bulletin of the International Society for the History and Bibliography of Herpetology **5** (1): 9–17.

Bloch, M. E. 1791. Naturgeschichte der ausländischen Fische. Berlin. Volume 5. viii + 152 pp., pls. 253–288. [not seen].

Blumenbach, J. F. 1799. Handbuch der Naturgeschichte. Göttingen. Volume 6. xvi + 708 + [32] pp., 2 pls. [not seen]

Blumenbach, J. F. 1800. Über das Schnabelthier (*Ornithorhynchus paradoxus*) ein neuentdecktes Geschlecht von Säugthieren des fünften welttheils. Magazin für den Neuesten Zustand der Naturkunde **2**: 205–214. [not seen]

Born, I. 1778. *Index rerum naturalium Musei Vindobonensis. Pars I. Testacea*. Vienna. [xlii] + 458 + [82] pp., 1 Pl.

Brown, T. 1816. The elements of conchology; or natural history of shells according to the Linnean System, with observations on modern arrangements. Lackington, Allen, & Co., London. [vi] + 168 + [1], 9 pls.

Bruguière, J. G. 1789–92. Histoire Naturelle des Vers. Encyclopedié Méthodique **1**(1): i–xviii, 1–344, 1789; **1** (2): 345–757, 1792.

Bullock. W. 1810. A companion to Mr. Bullock's Museum, etc., 8th Edition. Privately printed, London. iv + 98 pp.

Bullock. W. 1810. A companion to Mr. Bullock's Museum, etc., 9th Edition. Privately printed, London. v + 147 pp.

Bullock. W. 1811. A companion to Mr. Bullock's Museum, etc., 10th Edition. Privately printed, London. vi + 150 pp.

Bullock. W. 1812. A companion to Mr. Bullock's London Museum and Pantherion; etc., 12th Edition. Privately printed, London. xii + 57 + 136 pp., 31 pls.

Bullock, W. 1819. Catalogue of the Roman Gallery of Works of Art, and the London Museum of Natural History, which will commence selling by Auction, on Thursday, April 29, 1819, etc. Privately printed, London. v + 166 + [8] + 22 + [4] pp., 1 folding Pl. [consulted from the facsimile reprint with manuscript prices and buyers' names, published by Harmer Johnson & John Hewett, London, 1979].

Callery, B. G. 1981. The persistence of ephemera: bibliographic value of part wrappers. History in the Service of Systematics, Society for the Bibliography of Natural History Special Publication **1**: 15–22.

Chemnitz, J. H. 1777. Von einer ausserordentlich seltenen Art walzenförmiger Tuten oder Kegelschnekken, welche den Namen Gloria maris führt. Beschäftigungen der Berlinischen Gesellschaft Naturforschender Freunde, **3**: 321–331.

Cossmann, M. 1889. Catalogue illustré des coquilles fossiles de l'Eocene des environs de Paris, 4. Annales de la Société royale Malacologique de Belgique **24**: 7–385, 11 pls.

Cramer, P. 1775. De uitlandische Kapellen voorkomende in de drie Waereld-Deelen Asia, Africa en America. Papilons exotiques des trois parties du monde l'Asie, l'Afrique et l'Amérique. Amsterdam & Utrecht. Vol. 1 (1/7), xxx + 16 + 132 pp., 84 pls. [not seen]

Cuvier, G. 1800–05. Leçons d'Anatomie Compareé. Paris. 5 volumes.

Dall, W. H. 1904. An historical and systematic review of the frog-shells and tritons. Smithsonian Miscellaneous Collections **47**: 114–144.

Donovan, E. 1800. An epitome of the natural history of the insects of India, and the islands in the Indian Sea, &c. London. 57 pls. with descriptive letterpress. [not seen]

Donovan, E. 1805. An epitome of the natural history of the insects of New Holland, New Zealand, New Guinea, Otaheite, and other islands in the Indian, Southern, and Pacific Oceans, &c. London. 41 pls. with descriptive letterpress. [not seen]

Erxleben, J.C.P. 1777. *Systema regni animalis per classes, ordines, genera, species, varietates, cum synonymia et historia animalium. Classis I. Mammalia.* Weygandianis, Lipsiae. 636 pp. [not seen]

Eschmeyer, W. N. (Ed.) 1998. Catalog of Fishes. California Academy of Sciences, Center for Biodiversity Research and Information, Special Publication No. 1. 3 volumes. 2,905 pp.

Eschmeyer, W. N. (Ed.) 2007. Catalog of Fishes. http://www.calacademy.org/research/ichthyology/catalog/. Last accessed June 23, 2007.

Exeter Working Papers in British Book Trade History. 2005. http://www.devon.gov.uk/etched?IXP=1&IXR=100154. Last accessed September 21, 2005.

Fabricius, J. C. 1775. *Systema entomologiae, sistens insectorum classes, ordines, genera, species, adjectis synonymis, locis, descriptionibus, iobservationibus.* Korte, Lipsiae. [iv] + [xii] + [xvi] + 832 pp. [not seen]

Férussac, [A. É. J. P. J. F. d'A.] Baron de. 1820. Concordance systématique pour les mollusques terrestres et fluviatiles de la Grande-Bretagne; avec un aperçu des travaux modernes des savans anglais sur les mollusques. Journal de physique, de chimie, d'histoire naturelle et des arts: avec des planches en taille-douce **90**: 212–217 (January); 281–302 (March).

Férussac, [A. É. J. P. J. F. d'A.] Baron de. 1822. Tableaux systématiques des Animaux mollusques . . . terrestres ou fluviatiles, vivants ou fossiles. Paris. xlvii + 27 + 114 pp.

Finney, C. M. 1984. To sail beyond the sunset. Natural history in Australia 1699–1829. Rigny Publishers, Adelaide. x + 206 pp.

Fletcher, D. S. & Nye, I. W. B. 1982. In: Nye, I. W. B. (Ed.), The generic names of moths of the world. Volume 4. British Museum (Natural History), London. xiv + 192 pp.

Geoffroy, É. 1803. Note sur un nouveau mammifère décovert à la Nouvelle Hollande, par M. Bass, voyageur anglais. Bulletin de la Société Philomathique de Paris **3** (72): 185–1 86. [not seen]

Gmelin, J. F. 1791. *Caroli a Linné Systema Naturae per regna tria naturae. Editio decima tertia.* Georg Emanuel Beer, Lipsiae. 1 (6), Vermes, 3021–3910.

Goldfuss, G. A. 1817–24. Die Säugthiere in Abbildungen nach der Natur, mit Beschreibungen. Leipzig. Vol. 5, pt. 65–69. [not seen]

Gould, S. J. 1977. Ever Since Darwin. W. W. Norton, New York. 285 pp.

Gray, J. E. 1847. List of the genera of recent Mollusca, their synonyma and types. Proceedings of the Zoological Society of London for 1847: 129–219.

Gronow, L. T. 1772. *Animalium rariorum fasciculus. Pisces.* Acta Helvetica **7**: 43–52, pls. 2–3. [not seen]

Hancocks, D. 2001. A different nature: the paradoxical world of zoos and their uncertain future. University of California Press, Berkeley. xxii + 280 pp.

Illiger, J. K. W. 1801. Neue Insekten. Magazin für Insektenkunde, **1**(1/2): 163–208. [not seen]

International Commission on Zoological Nomenclature. 1967. Opinion 828. *Zorilla* I. Geoffroy, 1826 (Mammalia): Suppressed under the Plenary Powers. Bulletin of Zoological Nomenclature **24** (3): 153–154.

International Commission on Zoological Nomenclature. 1970. Opinion 911. Six misidentified type-species in the superfamily Muricacea (Gastropoda). Bulletin of Zoological Nomenclature **27** (1): 20–26.

International Commission on Zoological Nomenclature. 1992. Opinion 1677. *Haustator* Montfort, 1810 (Mollusca, Gastropoda): conserved. Bulletin of Zoological Nomenclature **49** (2): 159.

International Commission on Zoological Nomenclature. 1999. International Code of Zoological Nomenclature. Fourth edition. I.T.Z.N., London. xxix, 306 pp.

Iredale, T. 1915. A commentary on Suter's "Manual of the New Zealand Mollusca." Transactions of the New Zealand Institute **47**: 417–497.

Kerr, R. 1792. The Animal Kingdom, or zoological system of . . . C. Linnaeus: Class I. Mammalia, (Class II. The Birds) . . . being a translation of that part of the Systema Naturae, as . . . published . . . by Professor Gmelin . . . with numerous additions from more recent zoological writers . . . J. Murray & R. Faulder, London. xii + [xxviii] + 644 pp. [not seen]

Kohn, A. J. 1986. Type specimens and identity of the described species of *Conus* VII. The species described 1810–1820. Zoological Journal of the Linnean Society **86**: 1–41.

Lacepède, B. G. E. 1804. Mémoire sur plusieurs animaux de la Nouvelle Hollande dont la description n'a pas encore été publiée. Annales du Muséum d'Histoire naturelle **5**: 184–211, 4 pls. [not seen].

Lamarck, [J. B. P. A.] 1803. Mémoires sur les fossiles des environs de Paris. Suite. Annales du Muséum d'Histoire naturelle **2**: 163–169.

Lamarck, [J. B. P. A.] 1804. Mémoires sur les fossiles des environs de Paris. Suite. Annales du Muséum d'Histoire naturelle **4**: 212–222.

[Lightfoot, J.] 1786. A catalogue of the Portland Museum, lately the property of the Duchess Dowager of Portland, deceased, which will be sold at auction, by Mr. Skinner and Co. [London]. viii + 194 pp.

Link, H. F. 1807. Beschreibung der Naturalien-Sammlung der Universität zu Rostock **2**: 82–100; **3**: 101–160; **4**: 5–23.

Linnaeus, C. 1758. *Systema naturae per regna tria naturae. Editio decima, reformata. Vol. 1, Regnum animale.* Stockholm. 824 pp.

Mathews, G. M. & Iredale, T. 1912. "Perry's *Arcana*" – An overlooked work. Victorian Naturalist **29**: 7–16.

Mitchell, S. L. 1815. The fishes of New York described and arranged. Transactions of the Philosophical Society of New York **1**: 355–492, pls. 1–6.

Montfort, D. de 1810. Conchyliologie systématique, et classification méthodique des coquilles. Tome 2. F. Schoell, Paris. 676 p. (pre-28 May 1810; see Iredale, 1915: 457).

Mortensen, T. 1943. A Monograph of the Echinoidea III. 3. Camarodonta II. Echinidae, Strongylocentrotidae, Parasaleniidae, Echinometridae. C. A. Reitzel, Copenhagen. 446 pp.

Moyal, A. 1986. A bright & savage land: scientists in colonial Australia. Collins, Sydney. 192 pp.

Neave, S. A. 1939–40. *Nomenclator Zoologicus.* Zoological Society of London, London. 4 volumes: 1, xiv + 957 pp, 1939; 2, 1025 pp, 1939; 3, 1065 pp, 1940; 4, 758 pp, 1940.

Pallas, P. S. 1766. *Miscellanea Zoologica, quibus novae imprimis atque obscurae animalium species describuntur et observationibus iconibusque illustrantur.* Petrum van Cleef, Hagae Comitum. xii + 224 pp., 14 pls. [not seen]

Pallas, P. S. 1774. *Spicilegia Zoologica, quibus novae imprimis et obscurae animalium species iconibus, descriptionibus atque commentariis illustrantur.* Berolini. Fasc. 10: 1–41, pls. 1–4. [not seen]

Parenti, L. R. & Randall, J. E. 2000. An annotated checklist of the species of the Labroid fish families Labridae and Scaridae. J. L. B. Smith Institute of Ichthyology, Ichthyological Bulletin **68**: 1–97.

Patten, R. L. 2004–05a. Cruikshank, George (1792–1878). Oxford Dictionary of National Biography. Oxford University Press, 2004 [http://oxforddnb.com/view/article/6843, accessed 23 September 2005].

Patten, R. L. 2004–05b. Smeeton, George (fl. 1800–1828). Oxford Dictionary of National Biography. Oxford University Press, 2004 [http://oxforddnb.com/view/article/25751, accessed 24 September 2005].

Perry, G. 1810–11. Arcana; or the Museum of Natural History: containing the most recent discovered objects. James Stratford, London. 84 pls. with unnumbered text [issued in parts, pls. 1–48 in 1810, pls. 49–84 in 1811.]

Perry, G. 1811. Conchology, or the natural history of shells; containing a new arrangement of the genera and species. W. Miller, London. 4 + [61] + [1] pp., 61 pls.

Petit, R. E. 2003. George Perry's molluscan taxa and notes on the editions of his Conchology of 1811. Zootaxa **377**: 1–72.

Petit, R. E. & Le Renard, J. 1990. George Perry's fossil molluscan taxa, published in the 'Arcana' (1810–1811). Contributions to Tertiary and Quaternary Geology **27** (1): 27–35, pls. 1–3.

Röding, P. F. 1798. *Museum Boltenianum sive Catalogus cimeliorum e tribus regnis naturae quae olim collegerat Joa. Fried. Bolten, M.D.p.d.* Johan. Christi. Trappii, Hamburgi. 199 pp.

Rowe, F. W. E. & Gate, J. 1995. Echinodermata. In: Wells, A. (editor), Zoological Catalogue of Australia. Vol. 33. CSIRO, Melbourne. xiii + 510 pp.

Salis Marschlins, H. U. von 1793. Reisen in verschiedenen Provinzen des Königsreichs Neapel. Ziegler, Zürich & Leipzig [not seen]

Schreber, J. C. D. 1775–92. Die Säugthiere in Abbildungen nach der Natur, mit Beschreibun-gen. Erlangen. 5 vols., 1,112 pp., 347 pls. [not seen]

Shaw, G. 1790. *Macropus giganteus*. The Naturalist's Miscellany **1**: text to Pl. 33. [not seen]

Shaw, G. 1792. *Myrmecophaga aculeata*. The Porcupine Anteater. The Naturalist's Miscellany **3** (36): [6pp.], Pl. 109. [not seen]

Shaw, G. 1794. Zoology of New Holland. London. 33 pp., 12 pls. [not seen]

Shaw, G. 1799. *Platypus anatinus*. The Duck-Billed Platypus. The Naturalist's Miscellany, **10** (118): [7] pp., 2 pls. [not seen]

Shaw, G. 1800. General zoology or systematic natural history . . . Quadrupeds. G. Kearsley, London. Vol. 1, pt. 2. [not seen]

Shaw, G. 1804. General zoology or systematic natural history . . . Pisces. G. Kearsley, London. Vol. 5. [not seen]

Sherborn, C. D. 1902. *Index Animalium sive index nominum quae ab A.D.* MDCCLVIII *generibus et specibus animalium imposita sunt . . . Sectio Prima.* University Press, Cambridge. lx + 1,195 pp.

Sherborn, C. D. 1922–32 *Index Animalium sive index nominum quae ab A.D.* MDCCLVIII *gen-eribus et specibus animalium imposita sunt. Sectio Secunda. A kalendris Ianuariis,* MDCC–CI *usque ad finem Decembris,* MDCCCL. British Museum (Natural History), London. cxlvii + 7,056 + 114 pp.

Smithers, H. 1825. Liverpool, its commerce, statistics, and institutions; with a history of the cotton trade. Thos. Kaye, Liverpool. viii + 461 pp., 2 pls.

Swann, C. K. 1972. Natural history bookselling. Journal of the Society for the Bibliography of Natural History **6** (2): 118–126.

Sweet, J. M. 1970. William Bullock's collection and the University of Edinburgh, 1819. Annals of Science **26** (1): 23–32.

Torrens, H. S. 2002. The Practice of British Geology, 1750–1850. Ashgate Publishing Lim-ited, Hampshire. xiv + 356 pp.

Walton, D. W. (Ed.) 1988. Zoological Catalogue of Australia. Vol. 5. Mammalia. Australian Government Publishing Service, Canberra. x + 274 pp.

Whiston, W. 1696. A New Theory of the Earth, From its Original, to the Consummation of All Things, Where the Creation of the World in Six Days, the Universal Deluge, And the General Conflagration, As laid down in the Holy Scriptures, Are Shewn to be perfectly agreeable to Reason and Philosophy. London: Benjamin Took. [not seen]

Whittell, H. M. 1954. The literature of Australian birds: a history and bibliography of Austra-lian ornithology. Paterson Brokensha, Perth. xii + 116 + 788 pp., 32 pls.

Wilson, D. E. & Reeder, D. M. (Eds.) 1993. Mammal species of the world: a taxonomic and geographic reference. 2nd edition. Smithsonian Institution Press, Washington. xviii + 1,207 pp.

Wood, C. A. 1931. An introduction to the literature of vertebrate zoology; based chiefly on the titles in the Blacker Library of Zoology, the Emma Shearer Wood Library of Ornithol-ogy, the Bibliotheca Oslerianas and other libraries of McGill University, Montreal. Oxford University Press, London. xix + 643 pp.

PERRY'S
Arcana

FACSIMILE EDITION

ARCANA;

OR

THE MUSEUM OF NATURAL HISTORY:

containing the most

RECENT DISCOVERED OBJECTS.

EMBELLISHED WITH COLOURED PLATES,

AND

CORRESPONDING DESCRIPTIONS;

WITH

𝔈𝔵𝔱𝔯𝔞𝔠𝔱𝔰 𝔯𝔢𝔩𝔞𝔱𝔦𝔫𝔤 𝔱𝔬 𝔄𝔫𝔦𝔪𝔞𝔩𝔰,

AND

REMARKS OF CELEBRATED TRAVELLERS;

COMBINING A

GENERAL SURVEY OF NATURE.

✦✦✦✦✦✦✦✦

VOL. I.

✦✦✦✦✦✦✦✦

LONDON:

Printed by GEORGE SMEETON, St. Martin's Lane,

FOR JAMES STRATFORD, 112, HOLBORN HILL.

1811.

[This page was blank in the original]

INTRODUCTION.

T HE subject matter will chiefly consist of such objects in Natural History as have not hitherto been published, being mostly from countries which have Animals, Fish, Birds, Insects, and Plants quite new, with other interesting particulars; which can now only be found in large books that might be much too expensive for the general Reader. Natural History is become a much more general, and indeed delightful study, as it opens to our minds a strong connection which it has with all the Arts and Sciences; furnishing abundance of subject for the choice of the Painter, of Plants for the Horticulturist, of Medicines for the Physician, and of Minerals for the Chemist. To clear the pathway therefore to these studies, by an exact application of Drawings, accompanied by the Generic terms which are used for their distinction, must be highly useful to all enquiring Readers. To return however to the subject of the Arcana, it will not consist entirely of rarities seldom to be seen: sometimes representations will be given of Animals more common and well known, and generally in order to a stronger comparison of their forms and habits. The suitable colour will also be imparted to each, and by this means they will be recognized from things of the same form and of a distinct species. Such are the views and advantages of the Arcana, which it is to be hoped will draw down the patronage of the Public, since the plates will be minutely executed and the labour of all the parts fully attended to.

INTRODUCTION.

When we consider the expences of such a work, the increase of the prices in all the necessary costs, it is only to the growing accruement of the numbers sold, that the profits can possibly arise. To enliven the subjects of systematic description, naturally attached to the plates, the Editors will select from the remarks of different English travellers, and in their own words, such observations as were appropriate on the Climates and productions of different parts of the Earth which they visited. These they intend to publish as an Appendix to each Number, and thus afford a seasonable relief to the learned words and long sentences; which nevertheless are almost unavoidable. It is necessary perhaps in this part to abridge what we have to say of ourselves, lest our subjects should savour too much of Egotism. The Public will be the final judge of its worth, and no doubt decide according to their private or general opinion in the exhibition of the work to their candid criticism.

Several Naturalists of the present time have given to their readers, exaggerated accounts: such are the dreams of those who took many narrations upon trust, or added to them from their own fanciful imagination. Such as Bruce, Vaillant, and particularly Humboldt; the Editors of the Arcana will pass by their lucubrations, which are trifles unworthy of any attention, and turn with pleasure to the more solid and approved writings of Lister, Pennant, Ray, and Linnæus.

TO

J. C. LETTSOM, ESQ. M. D.

FELLOW OF THE ROYAL SOCIETY.

THE PATRON OF THE LIBERAL SCIENCES,

AND OF NATURAL HISTORY.

IN GRATITUDE AND RESPECT FOR THE NUMEROUS PROOFS

OF UNLIMITED KINDNESS AND REGARD:

AND IN REVERENCE OF THOSE EMINENT TALENTS

WHICH HAVE ALWAYS BEEN EXERTED

FOR THE GENERAL HAPPINESS OF MANKIND,

THIS SMALL TESTIMONY OF REGARD,

IS HUMBLY DEDICATED

BY HIS MOST OBEDIENT FRIEND,

AND OBLIGED HUMBLE SERVANT,

GEORGE PERRY.

[This page was blank in the original]

[This page was blank in the original]

Pl. 1

ZOOLOGY

J. C. Whichelo del.

T. I. Busby sculp.

T I G E R.

Published by J. Stratford. Holborn. Jan.1.ˢ 1810.

PLATE 1

ZOOLOGY.

THE TIGER.

Felis Tigris, Lin.   *Le Tigre*, Buffon.

AMONGST the various animals of prey which infest the sultry regions of Asia, there are few which do not yield to the Tiger in ferocity and strength. Driven from the more civilized haunts of Man, and separated from the society of domestic animals, he ranges through the silent and trackless forest, indulging his natural thirst for blood, dealing horror and devastation through the animal kingdom. In the hotter regions of the Globe where the smaller animals abound, and the Niger or the Ganges roll their tributary streams to the Sea, the Tiger reigns uncontrouled, and spreads his ravages amongst the numerous herds of Antelopes and Deer, where they resort to the springs or rivers for refreshment. If we could for a moment forget the power of his fangs, or the unrelenting fierceness of his nature, we might contemplate with pleasure, the beauty of his skin, and the elegant contrast every where

A

displayed in his form and contour. But the sense of his native cruelty, and irreclaimable nature, fill the mind with a secret and thrilling sense of detestation and horror, while astonishment usurps the place of pleasure.

The Royal Tiger of Indostan, and which is supposed to be of the largest race of these animals at present known, measures fourteen feet from its head to the end of its tail; his body is muscular and round, his feet large and projecting, armed with prehensile claws, each of them enclosed in a hollow horny sheath, like those of the Cat-tribe. His legs are short and not well calculated for swiftness, but rather for bounding or leaping upon his prey, for which he generally lies in wait, making a spring of twenty or thirty feet at a time upon the object he intends to seize. His tail is long and beautifully striped, in a similar manner with his back, having bands of dark brown placed across: and in this respect he differs materially from the Leopard and Panther, which are remarkable rather by their round spots scattered irregularly over their bodies.

It would be an astonishing circumstance to the human mind that the merciful Author of Nature should have created such animals only for the purposes of devastation, if we were not at the same time convinced how necessary it is, that the smaller race of animals should be reduced and kept under, and in this point the balance of nature is as admirably preserved, the fiercer and more powerful animals producing only a few young ones at a time. The Tiger notwithstanding his strength, has the peculiar cowardice never to attack his enemy in front, and unless urgently pressed by famine, it is probable he would not fail always to fly from man; but if assaulted, his rage gets the better of his fear, and he becomes resolute even to death. The Lion, Buffalo and Rhinoceros are his natural and formidable enemies, and with the Buffalo he is frequently enclosed by the Indian Chiefs in a stage-combat for the purposes of amusement, and in which case he generally becomes the victim.

Upon the whole there is reason to believe that the larger animals of prey, as the Lion, Tyger, and others, are much less numerous than formerly, as Europe is not able at present to exhibit any (except indeed in a captive state) although they formerly abounded there.

We subjoin the following description of the fight between the Buffalo and Tiger, as described by Captain Williamson in his Indian Sports. " A Pallisado is made of bamboo, thirty yards in diameter, and strongly fenced all round, from the top of which the spectators can behold the combat. As security is the soul of amusement, every precaution is taken to enclose the Area in such a manner as to obviate all reasonable fear. Where a Tiger is one of the Dramatis Personæ, too much care cannot be used, as there have been instances of their making their escape, and putting all the spectators to the rout. The walls of the Area are raised twenty feet high, and the populace are placed in an elevated gallery so as to command a view of the whole.

As soon as the Tiger has entered the Area, the gates are closed, and a short time is allowed him to look around and examine his new situation. At first he seems to creep in a cowardly manner close to the Pallisades, wishfully looking at the top, and grinding his teeth at the people who surround the Area. The Buffalo is then introduced, and nothing can surpass the animation displayed at this moment, the Buffalo's eyes sparkle with fury as he views his sculking enemy ; he rushes forward with his head down and horns direct, at the Tiger's body, which however serves rather to bruise him, than to tear his skin, which is smooth and pliant. The Tiger starts on one side and endeavours to plant himself upon the Buffalo's back by leaping over his head and neck, and in this he is often unsuccessful, and passing over him, changes his place and falling down, becomes submitted to the fury of his horns. The Buffalo however carries on a war of extermination, his rage being excited by his wounds, and the issue terminates uncertainly, but generally in the death of the Tiger, who becomes defeated through the greatness of the fatigue, and length of the combat. The violence of the Buffalo continues for some time after the fight; it is prudent therefore to leave him to cool, and to approach him with water and wet grass, of which he partakes with avidity. The road is afterwards cleared from passengers to prevent all accidents which might otherwise occur."

The present specimen was drawn from the beautiful Tiger in the Menagerie of Mr. PIDCOCK.

[This page was blank in the original]

Pl. 1

G. Perry del.

T. L. Busby sculp.

PALM TREE

Publish'd by J. Stratford, Holborn, Jan.ʸ 1ˢᵗ 1810

PLATE 2

BOTANY.

THE CEROXYLON, OR PALM TREE.

Polygamia.   *Monœcia.*

THE Ceroxylon, or Palm Tree of Peru, which has been submitted to the class of French Institute, by Monsieur Humbolt, is remarkable for it's novelty, as well as it's situation; for the lofty height to which it elevates it's summit; and the singular production of wax it yields; from which circumstance it has been sometimes called the Wax Palm.

Mutis, who has held a distinguished rank amongst modern naturalists, is the only one who had formed an idea of it's existence, which circumstance is mentioned in the supplement to the third Edition of Linnæus's Systema Naturæ.

According to the Botanical Distinctions of Linnæus, it must be classed with the Polygamia; of the Order Monœcia.

On the lofty and cloud-cap't summits of the Andes which separate the Valley of Madeleine from the River

Cauca, this tree chiefly abounds, amidst the most rugged precipices and barren passes of the country.

This Palm Tree is also a Native of Quindiu, one of the mountainous and snowy regions of Peru, and is called the Ceroxylon to distinguish it from the Palm Trees already known : it is said sometimes to reach the amazing height of 160 to 180 feet. The trunk is straight and swelling out in the middle, bearing at the top its immense branches in various directions. The fruit is small and round, containing an oval kernel; the flowers are of two sorts, growing out of a sheath ; the hermaphrodite and the female ; and are not remarkable for their beauty or their size.

The most extraordinary circumstance relating to this Tree, is the secretion of Wax, containing a small proportion of Rosin, through the whole outside surface of its bark, on each side of the circles where there have been the marks of the former leaves.

Pliny makes mention of a Larix Tree which was used in the Amphitheatre of Nero, and was 120 feet in heighth ; but the Tree at present under consideration, may be indeed regarded as the Monarch of all the Forests of the World, if its gigantic size can entitle it to that distinction.

No advantageous use has hitherto been made either of the Wax, which invests the bark of this Tree, or of the Fruit, both of which might it is supposed be converted to the uses of mankind ; the former for giving light ; the latter as a pleasant and wholesome food, and containing much sugar. The Timber is of a firm texture, and capable of being formed into beams and rafters for houses ; but the difficulty of removal from its original mountainous situation will perhaps be for ever an inseparable bar to its general use and consumption.

BOTANY.

The Palm Trees form a most astonishing family in the History of the Vegetable Kingdom; their amazing height; their majestic forms; the delightful and extensive shadows which they yield to the weary traveller, induced Linnæus to give them the name of Princes of India; and if we add to these external qualities, the Flour, the Wine, and the Oil, which they so plentifully produce, we may regard them as one amongst those Blessings for which Man has reason to be highly grateful to his Creator.

It is also supposed that this Plant might exist, by a comparison of the climate and temperature, if transplanted to the mountainous regions of Switzerland; hitherto however there is nothing more than conjecture to strengthen this opinion.

The Flower grows at the upper part of the Tree, shooting from a sheath or spatha, in clusters or bunches, upon which the berries are afterwards formed; the root consists of various arms and shoots spreading out at the foot, and giving security to the trunk.

The circular stripes which appear in the external bark of the trunk, indicate the gradual expansion of the Tree, each circle being formed every year, so that the relative age of the tree may be easily ascertained.

In the East Indies, the uses of the Palm Trees are extremely multifarious, for independently of the Canauca, Palm, which yields an excellent wax from it's leaves, by boiling; there is also another species, which supplies the natives with the following articles; bread, oil, milk, wine, ropes, masts, oars, cordage, clothing, wax, rosin, needles and thread.

BOTANY.

The Indians have a method of climbing these trees for the fruit, by placing two large hoops loosely round the trunk, into the lowest of which they place their legs, as far as the knee, and then raise themselves by the upper one, placing it at the utmost extent of their arms; at other times by shooting an arrow to which is fixed a rope, over the highest branches of the tree.

We do not find that the Seeds of this Tree have as yet been brought to Europe.

There has been a considerable difference in the opinion of Naturalists, as to the distinctive characters of the Palm Tree; some Botanists having proposed that these should be referred to the Genus Hexandria, and the Genus Polygamia wholly abolished. The Fig certainly differs so much from the Palms (by having its blossoms placed within the receptaculum), that it seems rather absurd to place them together; nevertheless as all artificial systems must be subject to some objections and contradictions, it seems better to leave the matter as it is laid down by the great Linnæus, than to abridge the number of the Genera, already specified, as Dr. THORNTON in his late Work has attempted, perhaps without sufficient reason. If any alteration were considered as adviseable in the Botanical system of the illustrious Swede, it would be better perhaps to enlarge than diminish the number of the Genera, as new discoveries of events are constantly made which do not readily reconcile themselves to the present established Genera.

[This page was blank in the original]

Pl. 1

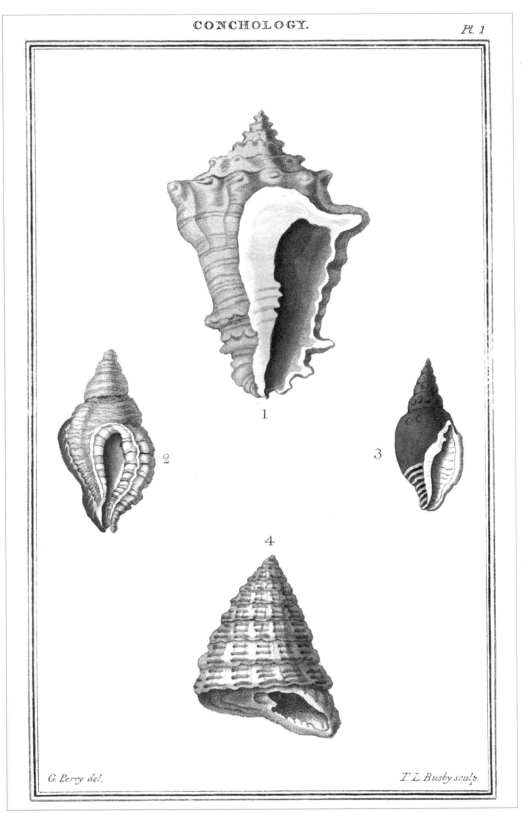

PLATE 3

CONCHOLOGY.

———————

THE variety and beautiful colours which are discoverable in the testaceous Family of Shells, have always rendered them an interesting subject to the Naturalist and the Man of Taste.

In describing the four Shells contained in the annexed Plate, we shall endeavour previously to explain the different Characters of each Genus, that the Reader may afterwards more easily recognize each peculiar Distinction appropriate thereto.

Shells have always been classed according to certain Similarities of Structure, observable in their outward form, and not from the qualities of the Animals contained in them, which are, generally speaking, quite unknown, except from Analogy to those which in the living state, are more easily within our reach.

In each Species, we shall elucidate the Genus to which it belongs, by its most striking peculiarities.

B

1. Genus, VOLUTELLA.

Character.—Shell univalve, spiral, the central pillar fluted with four flutes; the body and external cheek invested with tubercles irregularly placed.

Species.—Volutella divergens; Shell conical, angulated; of a bright yellow colour; the surface irregularly spinous all over; mouth oblong and labiated, of a rich pink colour.

This Species is very rare, is a Native of the Indian Ocean: *and is delineated from an Original Shell in the Collection of Mr. GREENHALL, of London.* The Genus is not very numerous, containing only about fifteen known Species.

2. Genus, SEPTA.

Character.—Shell univalve, spiral, having membranceous septa or divisions, placed upon the body and spire opposite and alternate; these are of a different colour to the rest of the Shell, and slightly tuberculated.

Species.—Septa scarlatina; Shell small, one inch and a half long; striped with scarlet bands, upon a yellow ground; the mouth white, verging to a brown colour.

This beautiful little Shell was formerly placed erroneously with the Genus Buccinum: it is a Native of Amboyna, in the East Indies, and varies from itself sometimes in having the colours very pale: It has been called by the Germans the Liveryhorn. *From a Shell in the Collection of the late Mr. WILLSON.*

3. Genus, ROSTELLARIA.

Character.—Shell univalve, spiral, having the outer cheek expanded (and united at the top of the mouth) to the spire; the beak straight and plain, ending in a point.

Species.—Rostellaria rubicunda; Shell ovated and slightly tuberculated; the mouth brown and striped; the spire and body of a dark red colour.

Like the former Shell it is a Native of Amboyna; and is from the curious and interesting Museum of Lord VALENTIA.

4. Genus, TROCHUS.

Character.—Shell pyramidically shaped, spiral; having the mouth placed underneath, leaning sideways, and of a quadrangular form; the spire inclined to the base.

Species.—Trochus Apiaria; Shell white, striped with green transversely and irregularly; the sides and base slightly rounded and tuberculated.

This curious Shell is a Non-descript, and lately imported from Botany Bay, a country which has afforded an ample field of new subjects for the Conchologist. *From a Specimen in the Museum of Dr. LETTSOM.*

CONCHOLOGY.

The general divisions of Testacea or Shell Animals, may be classed under the following Orders: Spirales, or Shells which have a twisted spire: Acuminatæ, or pointed Shells, as the Patella, &c.: Bivalvæ, or double Shells, as the Cockle, &c.: and lastly, Multiloculares, or Shells having bony compartments, as the Orthoceras, Nautilus, &c. Those which were denominated by Linnæus, Multi-alvæ, are not found upon Analysis to consist of similar component materials, and therefore ought properly to be separated; such are the Sabellæ, Chitons, &c. These latter are rather to be considered as Animals invested with a horney or membranaceous Covering, rather than Testaceous or Shell Fish; to which may be added, a very considerable difference in their internal Organization.

Philosophers have been much perplexed to account for the Manner of the growth of Shell-fish; and notwith-standing that matter has received a very copious Investiga-tion, it is still involved in considerable doubt. It was once believed that the animal had the power of adding an exter-nal Coat or Flap to the side of the Mouth, and which was repeated at certain intervals, enlarging the circle and size of the shell as the Animal increased in magnitude. Other writers have supposed that the Animal had the power of forming a new Covering for itself, and totally deserting the former shell when it became too small. It seems more pro-bable that the Shell has an internal Power of Growth or Expansion, which exists from its beginning or birth, and adapts itself by a general Expansion to the Size of the Animal. Certain it is, that when the Animal is arrived to its utmost size, it has the power of spreading over the whole Surface of its Mouth a substance of the smoothest Enamel, which serves at the same time to thicken and en-large the Lip; particularly the Strombus and Cowry.

CONCHOLOGY.

The Analogy which exists in the Vegetable and Animal Kingdoms is in many instances very striking and obvious. Amongst the Plants lately added by the recent Discoveries in the Southern Ocean, several have occurred which are *parasitical*, or living upon other Trees; of which we have a familiar instance in the Misletoe. The same circumstance occurs in the History of Shells, as several of the Lepas and Patella obviously shew, but the most singular Shell of this kind is the Proscenula, of which several have been lately discovered by Mr. STUCKBURY, in the Strombus and other large Shells, firmly adhering to the inside of the Mouth. This curious Genus which is hitherto undescribed, is flat and dish-shaped, in its general Form resembling the Patella, but differs from it in having its Apex or pointed summit placed at one end, and also below, a Proscenium, Platform or small Stage, projecting in a circular form, from under the Apex. Some of these are so small as to require the Microscope to investigate them fully; and indeed it is not at all improbable, but that the Number of those Shells which are concealed from our View by their smallness, is greater by far than those which are so obvious and fill our cabinets. Several Genera exist amongst the more minute kinds which are astonishing for the Singularity of their Forms, and the Beauty of their Colours. In this hitherto unfrequented Path we have only the Labours of two eminent Authors to guide our Steps; we allude to Fichtel and Soldanus, who have each of them scientifically endeavoured to sketch an imperfect outline for the arrangement of future writers. In the laborious descriptions of Mr. BOYS, of Sandwich, who endeavoured to form a general Account of some of the minute English Shells, the plates which accompany the Work are not either sufficiently expressive or beautiful. An hiatus therefore is left in this Part of Natural History, which by an accurate delineation of the objects may prove highly useful and entertaining. The Fossil Shells, which are found enclosed in the substance of our most solid Mountains (a lasting evidence of the general Deluge) hold out to the Naturalist a pleasing and interesting Field for Enquiry. Their Forms are so different to those found recent; their beautiful State of Preservation, and curious circumstance of their Enclosure in Beds of Rock or Clay, lay a claim to farther Enquiry and Investigation.

It is our intention to profit by these Remarks, and to bring forward from time to time the most singular and rare Species which may offer themselves to our observation.

The greatest number or portion of Shells at present known revolve spirally from the left to the right; but the Genera Helix, Melania, and Bulimus, are a remarkable exception to this rule, having a great many species which are reversed, still however even in this respect, varying sometimes to the right and sometimes to the left. The Diogenes Crab frequently for the purpose of safety and security, takes up his habitation in the deserted shell of some Whelk or Murex, and by this means furnishes a curious instance of natural instinct; from this time, one of his prehensile claws becomes gradually much larger than the other; that which is enclosed in the covering, shrinks up, and becomes useless, thus adapting itself admirably to it's newly-acquired situation. The Argonauta by the use of it's oars and sail, has particularly attracted the regard, even of the most ancient writers, and is supposed to have furnished the first idea of a ship. Different in it's internal structure is the form of the Nautilus, which has a regular assortment of chambered compartments, connected with each other, and which are entirely occupied by the animal. In short, the facts and observations which Conchology brings to our view, open to the mind, new scenes of continual admiration of that great Being, who has so wonderfully adapted their singular forms and instincts to the situations in which they are placed.

[This page was blank in the original]

ENTOMOLOGY.

Fulgora

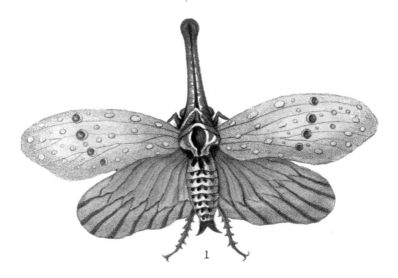

1

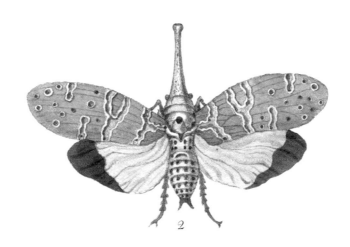

2

Perry del.

T.L.Busby sculp.

Published by J. Stratford Holborn Jan 1.st 1810.

PLATE 4

[84]

ENTOMOLOGY.

HEMIPTERA.

Genus, Fulgora, or Lantern Fly.

Character.—The Forehead truncated and rounded ; antenuæ underneath the eyes, and doubly articulated ; rostrum carved inwardly underneath.

No. I. THE FULGORA PYRORHYNCUS, OR BENGAL FIRE-FLY.

An Insect hitherto almost unknown and remarkable for the beautiful purple and green colour of its under wings. This singular animal which bears some general resemblance to the Genus Papilio, or Butterfly, has the extraordinary power of eliciting a Phosporic Light from the internal cavity of its trunk, which forms a striking character in its appearance—The wonderful power which the Glow-Worm possesses of illuminating by its small radiant lamp, the darkness of night, has been the theme of Poets, as well as Naturalists. The present Insect, which is a Native of Indostan, is endowed with a similar power, and contributes in no small degree to excite our wonder by the curious formation of its trunk or lantern, which is intended by Nature to light it on its way. One of the largest of this family, the Lanternaria, has been ably figured and described by Madame MERIAN, in her account of the Indian Insects. Having received from the Indians several of these, which she had carefully placed together in a transparent box, she was surprised in the night by their luminous appearance, and taking alarm at the display of the fire, as it indeed appeared to be, let fall the box to the ground, upon which the cause became obvious by the liberation of the Insects.

The upper wings of the present species are of a reddish brown, richly spotted; the trunk of a dark colour and rounded at the end; the under wings of a rich purple and green, alternately lanceolated in a pointed engrailment.

From a specimen in the Collection of Mr. SMITH, and is supposed to be very scarce; only two being at present known in England.

No. II. THE FULGORA CANDELARIA.

A Native of China, the trunk of a yellow colour turned upwards at the end and rounded; the upper wings green, streaked with beautiful veins of yellow; the under wings of yellow, edged with black. There is an agreeable contrast in the shades and tints of this beautiful Insect; but it is impossible to conceive what the effect of its light must be, except in its native Country, as it loses it phosporic effect when dried. Travellers who have visited China may be supposed to have exaggerated its effects, when they inform us, that the Indians perform their journies by night, carrying one of them fastened to the foot, and one in each hand, by this means making all other light unnecessary.

This Insect undoubtedly has light sufficient for its own purposes, the acquirement of its proper food, or the pursuit of its favourite mate; but of its uses to man we can form no such opinion, as Monsieur LESSER has figured forth in his *Theologia des Insectes*, who would persuade us, that the Natives use no other light in their houses, than this small phosporic animal.

The present Specimen is figured from the Original in the Museum of Mr. STUCHBURY, and exhibits the pristine colours in their full beauty and splendour.

[This page was blank in the original]

J.C. Wachelo del.

RATTLE SNAKE.

T.L. Busby sculp.

Published by J. Stratford Holborn Hill, Feb. 1st 1810.

PLATE 5

AMPHIBIA.

Crotalus horridus; or *banded Rattlesnake.* Linnæus.

Vipera, caudisona, Americana. Catesby.

THE Reptile, which is so well known by the dreadfully destructive power of it's poisonous Bite, is a native of the Western Hemisphere, and is found in most of the Islands, as well as on the Continent of America. It seems however to exist chiefly in the hottest parts only, and to be incapable of resisting the frigid state of all the colder regions. Three or four different Species are already known and described, of which the present Snake is the most important, as it frequently reaches the size of ten feet in length; and the stripes in the other kinds are much paler, and not transverse, but placed in a lozenge form along the back.

The distinction of the Serpents into the poisonous and harmless tribes, can only be known by an accurate examination of their teeth; the fangs or poisoning teeth being always of a tubular or channell'd structure, and calculated for the conveyance or injection of the poisonous fluid, from a peculiar reservoir, communicating on each side of the mouth.

D

The colour of the head of the Rattlesnake, is brown; the eye red; the upper part of the body of a brownish yellow, transversely marked with irregular broad black lists; the Rattle consisting of several horny membraneous cells, is brown and of an undulated form; the articulation of these parts being very loose, makes them rattle against each other when the reptile moves his tail, which he always does when irritated. Yet unless provoked or in pursuit of it's prey, the Rattlesnake must be considered as a sluggish and inactive animal, and is never the aggressor unless disturbed or assaulted. They make a deep wound and inject a very considerable portion of venom, but the poison if applied to the surface of the skin is said to be quite harmless, unless the skin be broken; and it seems to have no effect internally upon the stomach, as the Indians are in the constant habit of sucking the poison from the wounds of themselves or others; for this circumstance, we have the authority of Catesby and other writers on natural history; and it seems not improbable that the human saliva or spittle may be a true and natural antidote for the poison of almost all venomous Snakes, but which opinion however must lay open to farther experiments and observations. It is a remarkable circumstance also that the bite of venomous snakes should be fatal to themselves, which has been proved by experiments when they have been provoked in a state of confinement. The usual time of death ensuing after a person has been bitten by a Rattlesnake is from two to fifteen minutes; this however is supposed to depend upon the state of irritation of the animal at the time, and upon the constitution of the patient; there is reason to believe that in a state of captivity, it's operation would be weaker, as the Reptile is then generally considered as being out of health. The unfortunate man who was bitten to death lately in London, survived for 18 days, but in the greatest agony and pain, the inflammation being carried on to the greatest degree. The remedies applied were Salvolatile, Opium, and Brandy mixed with Water, but without the smallest good effect being obtained.

The Americans use a vegetable that abounds in their woods, and which is said to be a compleat and efficacious remedy, by the use of a decoction and cataplasm made from the root.

Of the Fascination or charming Power of the Rattlesnake, as it is commonly called, by which it can draw down from the highest trees small birds and animals, causing them to drop into it's very mouth, much has been asserted by travellers on each side of the question, the most probable explanation of this circumstance may perhaps arise from the terror and confusion into which the smaller animals are thrown by the sight and sound of the Rattlesnake preventing them from making their escape.

The Rattlesnake is a viviparous animal, producing it's young in the month of June, generally about twelve in number, and which by September acquire the length of twelve inches. It is said to practise the same extraordinary mode of preserving it's young from danger which is attributed to the Viper of Europe, viz. of receiving them into it's mouth, and swallowing them; they are afterwards seen to disgorge them when the danger is over. In Winter they retire to the most secret and inaccessable cliffs, where they pass the season in a torpid and dormant state, till the Spring once more brings them forth from their dark habitations.

Amongst the most remarkable circumstances in the conformation of the Rattlesnake, may be considered the mechanical Construction of his Tail. The form of it is so singular, that no description or even delineation of it will be found inadequate to explain its structure compleatly, without perhaps an examination. At the end of the body of the Snake, there is fixed a membraneous appendage which may be called not improperly, the Radix or Root of the Rattle, it is flattish in shape as if compressed; and upon this are placed ten or twelve horny rings or circles, which are connected loosely, and on the motion of the animal rattle against each other, the sound much resembling that of a person playing a game with dice. The circular rings have a strong hold upon each by means of a tightened collar, so that they may recede or be compressed upon each other, by any motion of the tail; and to this impulse the noise is always owing.

As the Rattlesnake changes his skin every year, it is supposed there is a fresh growth in the joints of the tail, and which varies according to the age of the animal. We shall finish our account of this extraordinary Reptile, by a concluding reflection, upon the wisdom and goodness of the great Creator, who has not left mankind, and other creatures, subjected to the danger of its deadly bite, without a means of alarm so well calculated to enable them to avoid and escape the evil. This vengeful animal is also very much confined in it's province and range, there being no authority to justify us in supposing, that they are ever found in the Eastern parts of the globe; they remain confined to the hotter regions of the Western Hemisphere.

From a fine living Specimen in Mr. Kendrick's Menagarie, Piccadilly, London.

[This page was blank in the original]

SEPTA TRITONIA

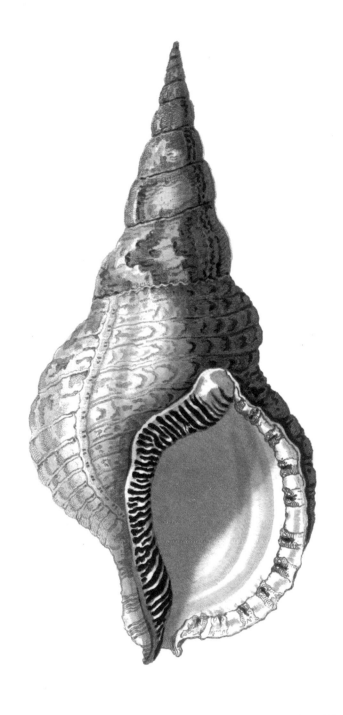

G.Perry del.   T.L.Busby sculp.

PLATE 6

Genus, SEPTA TRITONIA, or TRITON's HORN.

Character.—Shell univalve, spiral, acuminated, divided
   longitudinally by membranaceous sutures, placed
   irregularly and opposite, upon the folds of the
   spire, one of these forming the cheek of the
   Mouth or Maxilla Oris, the Columella or central
   Pillar corrugated or wrinkled. The Maxilla Oris
   is invested with double teeth painted, and of a
   brown colour.

THIS Shell, classed with the Genus Septa, and which
has hitherto been described erroneously as a Murex, is a
native of various parts of the Globe, being found in the
Eastern Ocean, and also in the European Seas. It is distinguished by the Richness of its Colours. It has sometimes been denominated the Triton's Horn, from the resemblance which it bears to some of the sculptured Relievos
of the Ancients, in which the Tritons, who wait upon
Neptune, are represented holding up Shells of this sort,
and blowing with them from their mouths a Music, suitable enough to those watry beings.

This remarkable Shell varies considerably in size,
being sometimes eighteen inches in length, and by making
a small opening at the upper end, a pleasing and agreeable
sound may be produced, resembling that of a trumpet, but
rather more deep and sonorous in its tone.

Another Shell, which has considerable resemblance in
its general form to the one now described, has lately been
discovered in New Holland, but it differs in the minuter

peculiarities of form and colour, being much smaller, and of a redder colour.

For want of proper and sufficient distinctions, several preceding Writers upon Conchology, have placed this Shell with the Murex Genus, but the Murex, strictly speaking, has no Divisions or Septæ on its Spire, in which may be instanced the Murex Morio and Murex Trapezium, &c. of Linnæus, which therefore must be always considered as belonging to a distinct Family.

From a charming Specimen in the Collection of Mr. Gwcnap, of London.

[This page was blank in the original]

J.C.Whichelo del.

CONDOR VULTURE.

Published by J Stratford. Holbern. Feb. 1810.

T.L.Busby sculp.

PLATE 7

ORNITHOLOGY.

THE CONDOR VULTURE.

Character.—Bill hooked, armed with a bulbous Base.
  Head and Neck partially bare, with a naked
    Skin.
  Neck curved and bent back.
  Feet armed with crooked Claws.

THE Vulture which has excited the miraculous and
fabulous narratives of those who have travelled through the
Regions of Southern America, has lately been introduced
into Europe in the live state, and is found by no means to
equal the astonishing Size which has been recorded of him.
The largest which has been known, did not exceed twelve
feet upon the extended wings. Nevertheless it so far exceeds the Eagle in grandeur and strength, that if size

alone were to constitute superiority, it might be truly denominated the King of Birds.

The Vultures in general differ from the Eagles in being of a heavier or less active character; in ferocity however and the untamable disposition of their nature, they are by no means inferior.

The Condor Vulture, the largest known at present is found only in South America, and has made its name terrible to the Natives by the attacks which it sometimes makes upon living animals, and in some cases even upon the human species. Some writers have confidently affirmed that it has been known to carry away Children where an opportunity has offered; and two of these birds have been seen to attack a full-grown heifer, and ultimately destroy it, by tearing it in pieces.

This curious Bird has a singular pouch placed under the lower mandible, of a blue colour, and reaching down the neck; it has also several fleshy appendages on each side of the throat, diminishing in size as they descend. Below the principal crest, which is large and upright, is a smaller one distinct and beset with coarse down. The crest is of a dark grey, and on the front of the neck is a pendent pearl-shaped tubercle; there is also a beautiful tippet of white fur forming an elegant collar round the neck, with the feathers turned back, and the claws are strongly hooked.

Since this bird was first exhibited in England, Monsieur Humboldt has published his Account of the Condor Americanus, and he mentions having frequently met with them on the Andes and Cordilleras Mountains in Peru. The young birds are entirely destitute of feathers, being covered with a fine whitish down, but which is full as thick as to give the young birds all the appearance of the old ones.

The Indians are in the habit of taking them by means of nooses prepared for them, and by means of baits of dead carcases sett for that purpose; for when the Condor has gorged itself with food, it becomes indolent and unwilling to fly, and is taken alive without much difficulty.

A curious stuffed specimen of the Condor Vulture, was lately preserved in the Leverian Museum, and afterwards said to be sold to the Emperor of Austria; we have no means therefore of comparing the measurement with the living Specimen, although from recollection, the size seems to have been much the same.

In its captive state it seems to have lost a great deal of its original fierceness, and to submit itself with a considerable gentleness of disposition to the different objects which surround it.

Birds of prey are said to have a greater longevity than others, and in this respect the life of the Condor Vulture is reported to coincide. The Golden Eagle has been said to have lived upwards of one hundred years, and Hawks and Falcons for a much longer term. Their affection for their young is very eminent, and at the times of hatching they are fearless of man and every external danger. Their nests are formed of sticks and dry grass, and are built upon the tops of the most inaccessible cliffs, amidst barren mountains, far from the peaceful and hospitable abodes of man, and where they can undisturbedly indulge in all the gloomy solitude of their nature.

The Condor Vulture, which is at present in Mr. KENDRICK's Menagerie, Piccadilly, London, may be re-

E

garded as a valuable acquisition to those amateurs who take delight in the curious parts of Natural History : it is stately and dignified in its appearance, and has preserved its natural appetite through all the horrors of transportation and imprisonment. He daily devours a large quantity of raw beef, and in appearance seems to preserve a vigorous and healthy constitution. In the annexed plate, he is described in the act of carrying away a native Peruvian Child; and as we have the authority of many grave and respectable writers to authenticate such a circumstance, we hope that we shall not incur the censure of the more incredulous and sceptical part of our Readers. It is no uncommon circumstance even in England for the Eagles to carry off Lambs of a considerable size; nor does it seem either extraordinary or improbable that a bird whose wings extend twelve feet from tip to tip, and whose conformation evidently marks him as a voracious creature, should if ever an opportunity occurs, readily and easily exercise its fatal powers upon the unprotected and helpless state of Infancy. The solitary nature of this bird however, and the particular regions to which he is confined, are providentially placed as a barrier and limitation to his otherwise boundless and voracious appetite.

[This page was blank in the original]

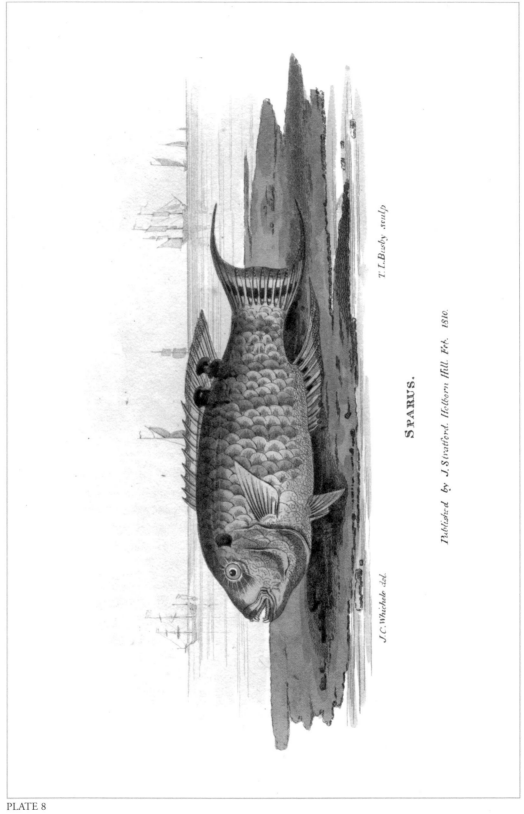

J.C.Whichelo del.

T. L.Busby sculp

S PARUS.

Published by J. Stratford. Holborn Hill. Feb. 1810.

PLATE 8

ICHTHYOLOGY.

Genus, SPARUS.

Character—Teeth strong, front teeth sometimes in a single, sometimes in numerous rows; grinders convex, smooth, and arranged like a pavement; lips thick; gill-covers unarmed; smooth scaly.

THE Sparus Bandatus is a native of the Eastern Ocean, and is distinguished by the elegance of its form and the richness of its colours, its eye is large and expressive, the tail very distinctly forked, the scales are semilunar and orbicrilated, having three bands of dark brown transversely placed upon the upper part of the back. It belongs to a very numerous Genus, of its peculiar habits therefore, nothing is very particularly known.

ICHTHYOLOGY.

REMARKS.

IN the various scales of living creatures which the extensive Field of Nature opens to our view, none is more calculated to strike us with astonishment and admiration than the Wonders of the Deep. Although the quantity of Birds, Animals and Insects which inhabit the terrestrial part of the Globe, is so great as almost to baffle all calculation, yet it is impossible not to suppose that the Ocean contains animals in a far greater variety and number than has been hitherto conceived or imagined.

In each individual Fish the increase of the progeny is almost incredible; if then we take into our enumerations the various Regions hitherto unexplored, the Bays and Gulphs, the Seas and Rivers, with all their boundless variety, we shall be lost in astonishment in so wide and extensive a view of Nature.

The habits and propensities of Fish are as various as their forms, whilst some by their voracious qualities are wisely designed to thin the over numerous swarms of the shallow coasts and rivers; others are singularly defended by a curious coat of external armour, resembling the spines of the hedgehog, or by a most deadly weapon fixed upon their beak, as the Sword Fish and the Narwhal.

The most destructive fish, the Shark, produces only a few young ones at a time, and by this means the admirable economy of nature is kept up, for if these fish were to be multiplied as rapidly as some of the smaller ones, the Ocean would be shortly exhausted of its population.

ICHTHYOLOGY.

But one of the most singular instruments of defence or annoyance imparted to fish is that which is displayed in the Gymnotus Electricus, or Electrical Eel, which has the power on being touched of imparting a violent electrical shock, and of repeating it for several times; and this quality seems to be inherited by several others, even after they are dead. By this means it is said frequently to fix upon its prey, and sometimes to defend itself from its pursuers.

> " Thus is nature's vesture wrought
> To instruct the wandring thought."

The Flounder, Soal and other flat fish which reside always at the bottom of rivers and bays, are deprived by nature of the air bladder, that natural instrument by which other fish can raise themselves to the surface of the sea.

The Whale, that fish of tremendous size, found chiefly in the Northern and Southern Regions of the Globe; is not found to be piscivorous, but exist chiefly upon a small kind of worm that exists in great abundance in those seas.

How astonishing is the instinct which directs the Herring at particular times of the Year, to seek a milder and more genial temperature, as well as the multitudinous swarms of Cod Fish which annually pay their expected visit to the seas and bays of Newfoundland.

Previous to a more particular description of the future species, the general character of this peculiar tribe of creatures will be more particularly pointed out. One of the most striking distinctions of their mode of life, is their constant residence and subsistence in Water, which is their natural and peculiar element, and for which they are admirably fitted by the gills or branchiæ, which answer the same purpose as lungs in other animals; without the intervention of an auricle and ventricle. The blood after being pushed forward by the heart into the ramification of the gills, is received by a large number of small veins which unite and form a descending aorta, as in the Mammalia tribe.

The general form of a fish may be not improperly compared to that of a Ship, the tail being regarded as the rudder, and the side fins as the oars, provided for impulse through the watery medium in which they dwell. For the purpose of raising or depressing their bodies in the Sea, they are provided with a curious air bladder, which by muscular compression can be made to condense the air contained therein, and by this means becoming themselves specifically lighter or heavier than the medium in which they move, they can easily rise or descend at pleasure.

Fishes are also endued with the sense of hearing, which has been made evident by several curious experiments; and the organ which is adapted for this purpose is situated immediately behind the eyes.

Their scales form a convenient kind of moveable armour, which is thoroughly covered with a glutinous substance for the purpose of gliding more easily through the waves.

They have also the power of smelling in a very exquisite degree, as is evidenced by their peculiar manner of taking or rejecting the bait.

Their eyes are placed variously, being most generally on the sides of the head, but on the flatter kinds of fish always at the top or summit, being in this manner more essential to their preservation.

The names of the fins which may not improperly be called their arms, have been distinguished in the following manner. The dorsal, or back-fins; the pectoral, or breast-fins; the ventral, or belly-fins; the anal, or vent-fins; and the caudal, or tail-fin. And lastly, the Cartilaginous fishes have only a membranaceous skeleton instead of a boney one, as their name naturally imports.

There are some kinds of Fish which have the singular property of being able to exist for a considerable time out of the water, as is the case with the Eel and the Flying Fish. It must be supposed that there is a corresponding difference in the organic conformation of the gills.

Fish are to be considered, by those conversant with mechanicle principles, as being admirably adapted by their form for the quickest and easiest transitions and motion. For this purpose they are shaped like a wedge, capable of cutting and dividing the medium through which they move, the nose being in general pointed, and the rest of the body gradually widening in breadth; this peculiar shape being the most favourable for swiftness of motion. Some species of Fish have their mouths placed under their head, as the Remora, or Sucking Fish, by which they can adhere very strongly to objects which they seize. By this circumstance the Shark is obliged to turn himself over, with his body up, before he can make his bite, and from this delay the life of the person devoted to his fury is sometimes saved.

The tail may be considered as a double fin, acting upwards and downwards; it can also impart a sudden motion forwards, in the manner of a scull or oar, such as is used by boatmen at the stern of their boats. The most surprizing efforts of all these, and depending upon the motion of the tail, is that which is exerted by the Salmon, in their passing over the celebrated Salmon Leap, at Ballyshannon in Ireland. The natural instinct of this Fish impells it at certain seasons of the year to resort upwards, through all the river streams, for the purpose of depositing its young, where it meets with a sudden and steep cataract or fall of water, its course would be ultimately stoped; the Fish however doubles its tail round as far as the mouth, and by a sudden and elastic expansion of the tail forces itself into the air; thus by repeated efforts gaining a greater height than the cataract, it at last regains the uppermost stream.

It is probable also that the Flying Fish throws itself into the air by a similar means and process, where it uses them for a long time as wings only, but afterwards when they become dry is obliged to drop down again to its native element.

The specimen of the Sparus Bandatus, was engraved from a beautiful correct drawing presented to the Editor; the accompanying embellishments by Mr. Whichello.

[This page was blank in the original]

J.C.Whichelo, del.

T.L.Busby, sculp.

NONPARIEL PARROT.

Published by J.Stratford, Holborn March 1.1810.

PLATE 9

PSITTACUS NONPAREIL; or NONPAREIL PARROT.

Character—Bill hooked, prehensile; feet scaled and strongly armed with claws; the head and neck scarlet;
the back blue streaked with yellow.

THE specimen here described is a native of Botany Bay,
and has lately been imported alive into England; its plumage consists of an assemblage of the richest and most
striking colours, and is delineated from the Museum of
Mr. BULLOCK. In size it is considerably less than the
common parrot, but does not resemble it in the imitation
of the human voice; the cry which it sometimes utters,
being rather like that of a Turtle Dove. The Parrot,
Parroquet and Lory differ chiefly from each other in the
size of the body, and in the form of the tail, but the
general discriminating character of the bill is similar
throughout the different tribes.

The celebrated naturalist, LINNÆUS, has divided the
families of the feather'd part of the creation into six orders;

1. *Accipitres;* or *Predacious Birds:* such as Vultures,
Eagles, Hawks, Owls, and a few others, distinguished by
the bill being of a crooked form.

2. *Passeres;* or *Passerine Birds:* comprising Pigeons,
Larks, Thrushes and all the Finches or small birds in general, either with thick or slender bills.

F

4. Gallinæ; or *Gallinaceous Birds;* or such as are more or less allied to the common domestic fowl, and consequently containing the Pheasant and Partridge-tribe, the Turkey, Peacock, and several other birds.

5. Grallæ; or *Waders:* consisting of all the Heron tribe, the Curlews, the Plovers, &c. having lengthened legs, and chiefly inhabiting watery situations.

6. Anseres; or *Web-footed Birds:* as the Swan, Goose and Duck tribes; the Gulls, Penguins and many others.

Out of these six Linnæan Orders, some Naturalists have instituted a few others in order to give a greater degree of precision to the arrangement; nevertheless it cannot be considered as absolutely necessary. Thus the Pigeons have been sometimes considered as properly forming a distinct order of Birds, under the title of Columbæ, or the Columbine Birds instead of being ranked among the Passeres of Linnæus. The Ostrich, Cassowary and Dodo have been supposed to constitute a division called the Struthious Order, instead of being placed with the Grallæ or Gallinæ of the former writer.

Birds are distinguished chiefly from other animals by the following singularities. In the circumstances of their anatomy they may be described according to the ancient method, as a two-footed, feather'd animal: the breast bone is solid and shaped like the keel of a ship, for the purposes of greater security and cleaving the air; the arms (as they would be called in other animals) are covered with long feathers, and answer to the design of Nature in their winged flight; the mouth is triangular and projecting; the tail spread out more or less in a feather'd extremity; the down which is frequently involved by the larger feathers, is of a soft texture like hair, and the quills of the wings gradually vary in their size from the origin to the extremity, and are capable of being folded up closely to the body; the feet are divided into toes or claws branching out and armed at the ends with a strong hook or point; but the most remarkable circumstance of all is their bill which answers the purpose of mouth and nose; the eyes are placed upon each side of the head, by which means they are more protected from external injury and are invested with a curious nictitating membrane, by which they can exclude any degree of light when found to be too powerful. The instinct of birds is no less surprizing than their structure, the conjugal attachments which they form, so necessary to the protection and support of their young, the long and fatiguing journies performed by the migratory tribes of birds, are proofs of Providence the most striking and decisive.

The beauty and splendid plumage of the trophical birds has been the general theme of admiration with almost all travellers. Nevertheless it is much to be questioned if they who reside in the temperate regions of the globe, would willingly exchange those feather'd songsters which charm them in every succeeding spring, for the gay Birds of Paradise or the splendid Macaws. The inhabitants of the hotter climates of the East and West Indies are frequently stunned and wearied with a continual noise which results from the vocal tenants of their forests. The Saw-Bird, so called from the incessant croaking noise which it makes in the night time, is enough to weary the most resolute patience with its monotonous sounds. In England, if we wish to be charmed with the songsters of the grove, it is always from choice, not from necessity, and we must court the lonely Philomel, if we wish to enjoy her enchanting and unobtrusive notes; this circumstance has been beautifully illustrated by MILTON:

> Thee chantress of the Woods among
> I woo to hear thy evening Song!

If the melodious qualities of song are to be highly valued in preference to gaudy colours, we possess the harmonious Wood-lark, the cheerful and sociable Robin Redbreast, the active and lively Sky-lark, to awake our senses to a delight for the charms of melody, and which breathe into our minds a more congenial sentiment than can possibly be derived from any foreign productions. As nature gives not all great qualities together, it is possible to admire the beauties of their plumage, whilst at the same time we lament their want of harmony and association to the human feelings.

The Parrotts are generally found in the hotter climates of the globe, and are distinguished by their crooked bill and the peculiar form of the claws. Those which have been lately discovered in New Holland form a numerous assemblage of new and striking characters hitherto undescribed. From these we have selected the Nonpareil Parrot, which for the richness of it's scarlet and blue plumage may be justly appreciated as one of the most beautiful of it's tribe. It's head and neck are of a deep scarlet; the back blue, striped with yellow; the bill and legs brown, and it's character is more lively and interesting than most of its congeners. Of its native habits however, we are at present little acquainted, as the attention of travellers in New Holland. has been so much arrested by the great variety of new objects as to prevent them hitherto from enquiring closely into the character of each individual species.

[This page was blank in the original]

Cruikshanks del.^t

T L Busby sculp.^t

PLATYPUS.

Publifhed by J Stratford Holborn March 1.1810.

PLATE 10

PLATYPUS; OR ORNITHORINXUS PARADOXUS.

THE singular structure and appearance of the animal which we are about to describe seem to remove it equally from almost every creature at present known, and with which, according to the Linnæan system, we should be inclined to class it. The extensive Continent of New Holland, or rather Island (as it may more properly be termed) being entirely surrounded by the Sea, is now ascertained to be of an amazing size, larger than the whole of Europe, and to contain animals of quite a different nature to those found in the other parts of the World.

This is to be considered as the strongest natural proof to a reasoning mind, that the Flood or Diluvian Overflux of the Ocean, was not universal, for if so, it would be impossible to account for the restoration of each individual species to each particular climate.

The Plants, the Insects, the Animals, and even the Fish, are in this new and lately discovered region, entirely distinct and secluded in their nature and manners; even man himself seems to differ here from his own species in the peculiarly untractable and savage constitution of his mind. Although it is not very improbable that the human species may have emigrated to this singularly detached country, from the neighbouring Islands of the South Sea, yet it is utterly incredible that the animals could have done so, or that they could have been brought there for any particular purpose, as they are not to be found any were else in a similar state. The Kangaroo, the Opossum and the Wombach of New Holland, and above all, the animal about to be described, are so evidently distinguished, that we must consider them as created in their present situation, as one great link in the chain of animated existences. Some modern philosophers anxious to account for the infinite varieties of animals and plants, found in different regions of the Globe, have asserted, that by the lapse of ages, or the change of food and climate, such an alteration may gradually take place, as to make from the same individual an apparently different species. Such a theory however seems by no means reconcilable with the generally acknowledged facts of nature.

The Platypus might be classed along with the Seals, if we were to consider only its external appearance, as its legs are very short and invested with a membranceous fin between the toes for the purposes of swimming, and which stamps its character as an aquatic animal. Its nose or bill much resembles that of a Duck; there are no teeth, but in place thereof is a serrated ridge on the internal edges of the under mandible. The length of the whole animal is thirteen inches, measuring from the tip of the beak to the end of the tail. It resides chiefly in watery situations on the banks of rivers, and its food is supposed to consist of aquatic plants and animals. On the upper part of the head, on each side, a little beyond the beak, are situated two small oval spots of white, in the lower part of which are imbedded the eyes, or at least those parts which Nature has allotted for vision, but they seem (perhaps like those of some of the Moles) but imperfectly calculated for distinct vision. Its general conformation appeared so extraordinary that when first dis-covered, some eminent Naturalists suspected an intention of deceit in the different descriptions given of it, but several specimens being lately obtained from New Holland, their doubts of its originality became compleatly removed.

A second animal of the same Genus, and which may be called the Platypus Longirostra, has lately been shot in

ZOOLOGY.

Adventure Bay, at Van Diemen's Land, and is supposed to be closely allied to the above in its external and internal habits. The chief difference consists in the tail being much shorter, and the nose much more taper (but still resembling a Duck's bill) and the body covered with a brown coat of thick hair interspersed thinly with blunt quills. It was 17 inches long and walked about two inches from the ground. The above description agrees with an accurate drawing made upon the spot at the time, and brought over to England by an eminent Naturalist. These two animals have been considered by Dr. SHAW, as having a very great analogy to the Myrmecophaga or Ant-Eater, which it resembles in the circumstance of being without teeth, but the feet certainly are very different, as also its ears, which consist merely of open uncovered foramina, and are placed directly behind the eyes. The feet of the Ant-Eater have separate claws, but those of the Platypus are united by a strong membrane, the distinguishing character of all animals which reside much in the water.

The back of the Platypus is covered all over with thick and close hair of a dark brown colour, much resembling that of a young Otter, but though obliged to walk very awkwardly upon land by means of the shortness of his legs, yet there is no doubt but that in the rivers, he can make a more rapid progress.

It is an observation which has not escaped the regard of those Naturalists who have described the creatures of New Holland, that all the quadrupeds hitherto discovered in that extensive region are void of symmetry in form and beauty of colours, whilst in the feather'd tribes, and in the vegetable kingdom the greatest profusion of beauty prevails. No beasts of prey have hitherto been discovered, a few species of the Racoon, Opossum and Kangaroo being all

[121]

the animals hitherto known, although it is much to be expected that when the vast internal forests of the country come to be explored, that animals may perhaps be found exceeding in singularity and in size any which are now delineated. Thus the veil of Nature will be gradually removed, and the advantages of Commerce and Science ultimately extended to the most distant and unknown regions of the Globe.

All the animals of New Holland seem to be formed with legs and arms either too short or too long. This is remarkably the case with the Kangaroo, the Platypus and with the Kaoli, a new animal of the Sloth kind, lately brought over from that Country. As this animal is entirely new and hitherto undescribed, it is our intention to give a delineation and description of it in one of our succeeding numbers, from the original animal in Mr. BULLOCK's Museum. The investigation of the different varieties of Nature is at least highly interesting and instructive, although not at all times reconcilable to our preconceived ideas of beauty or of general utility.

The Platypus seems wholly deserted by Nature, as to any means of defence from its enemies, or from animals of superior strength, and may therefore be considered as a perfectly harmless and timid creature.

[This page was blank in the original]

C.Blackelo.del.   T.L.Busby.sculp.

GREEN PARROTT.

Published by J.Stratford,Holborn,March 1.1810.

PLATE 11

[124]

PSITTACUS VIRIDIS. GREEN PARROQUET.

FROM NEW HOLLAND.

THE Aras, or Macaw, has been generally placed by Naturalists at the head of the numerous family of the Parrot tribe, and that chiefly from their superior size and the magnificent display which they make of their great length of tail. This distinction seems very proper and indeed due to them, as it is according to the order of other arrangements of Natural History, the Eagle and Vulture being placed before the Falcons and Hawks; by this means we naturally descend to the smaller and less conspicuous kinds of the Parrot and Parroquet. The Aras, or Macaw, is discriminated from the other orders by a very particular mark, which consists in the naked cheek, or rather by a naked membrane, which being without feathers, embraces not only the whole of the face but also the lower mandible of the beak. This membrane which surrounds the eye, gives to the Physiognomy of the Aras, a disdainful and disagreeable character, it is always found to be white in the Aras of the New Continent, at least in those species hitherto found. All of them have the tail very long and variously divided and joined; to these peculiar characters of all the Parrotts in general, a bill strong and crooked, which serves them for climbing; the upper mandible moveable, the tongue plump and round and quite blunt; the nostrils round and situated at the base of the beak, two toes before and two behind, of which the foremost are very much flattened; the tarsus of the foot is short and depressed, and which forms a rest for their feet when walking.

These birds, according to the report of travellers, generally fly in troops; they perch on the most elevated branches

G

of the forest trees; they feed on the different fruits, chiefly of the Date kind. They are docile and capable of being tamed and are easily taught some few words, but their tongue is too thick for them to speak distinctly; and with a strong and harsh voice they habitually repeat the word **Arra,** from which they take their name ; they are also long lived in their own country, but greatly susceptible of the impression of a colder atmosphere.

The Parroquets on the other hand are distinguished by a bill and face covered with feathers, and from the different form of their tails, they have been divided by LEVAILLANT into three families, of which a future and more particular description will hereafter be given when the separate species will be elucidated.

The present bird is the Green Spotted Parroquet of New Holland, and is supposed to be hitherto undescribed. It is delineated from a specimen in the Museum of Mr. BULLOCK, and is of a form and character highly pleasing. Its general colour is of a uniform grass green richly variegated and adorned with black angular spots, the hinder feathers of the wings brown, the bill black, and the tail-feathers long and spotted alternately with black and light green spots. Nature seems to sport with unbounded variety in the plumage of the Parrot tribe, yet the transition of the shades is generally so gradual, owing to the reflection of the rays that every harsh contrast seems to be carefully avoided. The tufted species are adorned in a remarkable manner by the spreading crest, which gives a singular appearance, as they have the power of raising or depressing it at their pleasure. The imitation of the human voice, which in some of them is so close, as to be hardly distinguished, adds much to the interest which they otherwise gain over Mankind, and in some instances they seem, in a certain degree, to possess that intelligent principle which is denominated Reason. The number of the species already discovered, it is supposed must amount to about three thousand, and when the inner parts of the countries near the South Pole, are farther investigated, there is little doubt but this curious part of Natural History will be still further increased and enlarged.

To the above general description of the Parrot tribe, we may add this singular circumstance respecting their mouths, namely, that they have the power of opening the mouth wider than any other bird, by means of an elongation of the hinge of the jaw, without which they would be unable to eat their food, owing to the great curvature of the upper mandible of the bill. Their feet are formed like those of the Cameleon, with two claws before and two behind, to enable them to ascend or descend with greater ease amongst the branches of the trees, also to hang downwards and turn round, of which practice they seem to be particularly fond.

The terms used in describing the Parrot, Parroquet, and Lory, have been indiscriminately used and confounded with each other, by which great confusion has been introduced. Some of our Naturalists following the example of LATHAM, have placed the crested Parrots in a distinct family, but there seems hardly a sufficient reason for so doing, for if two birds agree with each other in all respects, excepting the having a crest or having none, the Genera might then become too numerous for any convenient purposes of Classification. Several of the species of Birds which are crested, particularly the Grebes and Starlings, are not divided from their congeners, upon the small circumstance of a difference as above mentioned. The form and length of the Tail is indeed another strong mark; and it seems proper enough that the form of the Bill should be taken into consideration.

ORNITHOLOGY.

The French author, Monsieur LEVAILLANT, has in this respect, adopted, as we conceive, a very laudable and perspicuous arrangement, by placing the Aras at the head of the grand Work which he has lately published upon this most interesting subject, afterwards dividing the remainder of these Tribes into three Genera, by the distinguishing characters of the Tail. It is our intention therefore to adopt his system in the future descriptions in this Work, regarding it as more systematical and classical than any other that has hitherto been published.

For this purpose we shall shortly present to our Subscribers, a correct Representation of the Ara Militaris, or Military Macaw, from a fine specimen in the Collection of Mr. BULLOCK, recently brought over from the South Seas. These Birds, (the Aras,) partake very much of the character of the Eagle, and may be denominated the Kings of the Parrots, from their superior size and the dignity of their carriage and demeanour.

In a subsequent number, will also be given, an exact delineation of the Termes Bellicosus, or African White Ant, from the same valuable Museum before mentioned.

[This page was blank in the original]

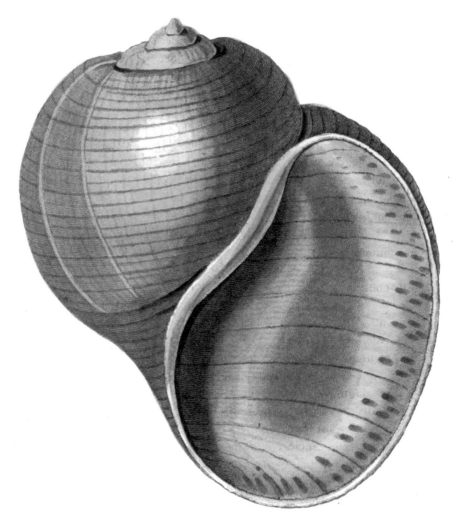

G. Perry, del.ᵗ   I.L.Bushy, sculp.ᵗ

POMACEA MACULATA.

Published by J. Stratford, Holborn March. 1 1810.

PLATE 12

POMACEA MACULATA.

Character—Shell univalve, orbicular, spire short, round and obtuse, mouth open and divided by a circular margin from the body, beak none.

THE Shell at present to be described, is analogous to the Helix or Snail in its form and appearance, and has generally been classed with that Genus by former authors, nevertheless its distinctions are sufficiently striking to have prevented such a gross error in its arangement, the mouth being divided all round from the body by an upright and distinct margin, which the Helix or Snail Shell is always without. It is therefore not unappropriately denominated the Pomacea or Apple-Shell, from its general resemblance to Pomum, an Apple, the Latin name for that well known and familiar fruit. It is delineated from a specimen in Mr. BULLOCK's Museum, and is conceived to be a Native of the South Sea, but of what part is not at present exactly known. It may certainly be considered as being very rare. The colour on the outside is of an olive green; its mouth of a pale brown, spotted with brown marks; the spire very small and short, but at the same time strongly furrowed and very distinguishable. All the fish which are to be found in this family of Shells, are highly delicious in their flavor, and form a most nourishing species of food. We are informed that the ancient Romans had so great a fondness for Snails, that they had wells constructed for the purpose of feeding them, and that they were afterwards sold at very considerable prices.

The Moderns seem to hold them much in contempt, and although constantly exposed for sale in the public markets, seem to be merely appropriated to the sickly and weak, being considered by some as a powerful restorative in cases of Consumption.

Several of the earliest writers upon Conchology, had divided the Shells of the Sea from those of the Land, by the distinguishing names of the Terrestrial and the Marine, but this division is now overlooked by the circumstance of several of them living alternately in fresh water, or Bays of the Ocean, or in Rivers where the Tides occasionally flow inwards and outwards.

The Genus Pomacea does not form a very numerous assemblage, and indeed has been most surprisingly overlooked by most of our recent authors. Very few have hitherto been found on the coasts of England, and those in general very small. Its natural place is the next in order to the Genus Helix, before mentioned; it approaches also in some of its characters to the Genus Bulimus, Melania, Ancilla, and several others which are without a beak, and also reside generally in the fresh water rivers and lakes of different regions.

ENTOMOLOGY.

An Account of the Termites Bellicosus, or White Ants, found in Africa: extracted from Mr. Smeathman's Travels in that Country.

THE curious history of the Termites, or White Ant of Africa, has attracted the notice and investigation of almost all travellers who have visited that immense country, but hitherto in a very imperfect and unsatisfactory manner. These extraordinary animals which erect for themselves buildings of clay, twelve feet high and generally about six feet broad at the base, are distinguished, like the Bee, the Ant and other social animals, for the singular art with which they construct their habitations, which are built with great strength and solidity. They appear to subsist chiefly upon decayed timber, or wooden posts found in the villages which are deserted by the natives, and of these they will devour an amazing quantity; the reproduction and multiplication of their own species being astonishing, rapid, and multifarious. Of the species called Bellicosus, there are three orders, the working insects, or Labourers; the fighting ones or Soldiers, which do no kind of labour; and lastly the winged ones, or perfect insects, which are male and female, and capable of propagation.

These last might very properly be called the nobility or gentry, for they neither labour, or toil, or fight, being quite incapable of either, or even of self-defence. These only are capable of being elected Kings and Queens, and Nature has so ordained it they generally emigrate in a few weeks after they are elevated to this estate, and either establish new kingdoms, or perish within a day or two. When these insects attack those things which man would not wish to be injured, they may be considered as being most pernicious, but when they are employed in destroying decayed trees and substances, which only encumber the surface of the earth,

they may be justly supposed very useful. The rapid vegetation in hot climates, of which no idea can be formed in
other countries, is equalled by as great a degree of destruction, from natural as well as accidental causes. So when
trees, and even woods, are in part destroyed by tornadoes or
fire, it is wonderful to observe how many agents are employed in hastening the total dissolution of the rest. In some
parts of Senegal, the number, magnitude and closeness of
their buildings, make them appear like the villages of the
natives, the form of each building being like a sugar loaf.
The inner part is divided into an amazing number of apartments, for the residence of the King and Queen, and is
considerably larger than the others, it being constantly
also in the centre of the building. The Queen, when at
her full size, becomes very large, and she, as well as the
King, can never possibly go out, as the entrances and passages are only just wide enough to admit the Soldiers or
Labourers, of which great numbers are necessary, and who
are always in the adjoining apartments, to which there are
numerous side passages communicating with each other.
Near these on each side are the magazines and nurseries; in
these are the provisions kept, which consist of raspings of
wood and the particular gums of different plants. There
are also several wide galleries, which intersect the building
in different directions, and the oven, or cell, which contains
the Queen, is placed on level with the external ground and
in the centre of the whole.

It appears that when these animals devour the posts and
beams of the roof of a house, they replace the cavities
which they make by a kind of clay, this, it is supposed, is
to prevent the Ants from following them, and KEMPFER
relates an instance of their piercing the leg of a table, then
passing on by the top and down the opposite leg, without
injuring several papers which were left upon it.

[This page was blank in the original]

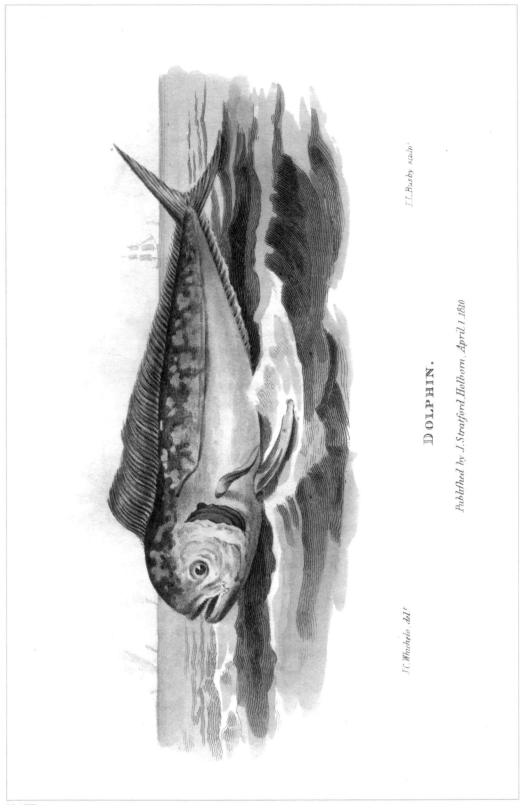

J.C.Whichelo. del.

DOLPHIN.

Published by J.Stratford, Holborn, April 1.1810.

J.L.Busby sculp.

PLATE 13

THE DOLPHIN.

Delphinus of Pliny. Porcus Marinus of Sibbald.

Character.—Body oblong, round, snout narrow and protuberant.

THE Dolphin bears a considerable resemblance in its external form to the Porpoise, but its nose is more elongated and acute, the shape of the body also is more slender throughout, it also grows to a much larger size, and it sometimes reaches to eight or ten feet in length. The colour when alive is said to be of a bright green, spotted with white, which changes much like the Mackarel, when it expires; it preys on various kinds of fish and is said to be sometimes seen attacking and wounding even the larger kinds of Whales. The mouth of the Dolphin is amply furnished in each jaw with a double row of teeth, and it may on the whole be considered as closely allied to the Shark-species. It is said to swim in a crooked posture, something in the way described by the ancients in their works of sculpture.

The Dolphin which is herewith delineated, is from the Museum of Mr. Bullock, and is about three feet in length; the nose of this fish is round and sharply projecting forwards, ending in a high ridge, continued into a long fin upon the back, the belly-fin is also very long and continuous, the colour of the back is a bright green, with white spots. Along each side of the body there runs a line of a dark green colour, which forms a pleasing ornament in the appearance of the fish, and the under jaw

H

of the mouth projects considerably beyond the upper part of the head, the form altogether is admirably calculated for swift sailing. Many fabulous reports have been circulated at different times, by former writers, of the attachment of the Dolphin to mankind, of their following close to the vessels, and sporting in a wanton mood to the sound of music, as if they had taken particular delight in the sound of instruments. The poet Ovid describes the excellent musician Arion, as having performed so admirably on the Lyre, that he was carried on the back of a Dolphin, safe to land, in a situation where the rest of the mariners were inevitably lost and shipwrecked. But these devices are to be considered only as the lawful fiction of the poets, who always delight to deal in the marvellous. One circumstance however, is indeed very remarkable, respecting the Natural History of the Dolphin, and which is strongly confirmed by all navigators, which is the singular occurrence of a change of colour, which takes place when removed from its native element, the the whole body becoming of a bright pink colour, previous to its death.

The number of species is not at present accurately known, but it is reported that in the Atlantic Ocean several kinds exist of a size much superior to those found in the European seas, but which are difficult to preserve, from their immense size and consequent tendency to putrefaction.

[This page was blank in the original]

PLATE 14

Cruikshanks delt.

VICUNA.

Published by J Stratford Holborn April 1st 1810

T.L.Busby sculp.

THE VICUNA.

Camelus Pacos, Linnæus. La Vigogne, Buffon.

THE Pacos or Vicuna of South America, is one of those animals which was formerly used by the Indians as a beast of burden, as well as the Lama, which is a larger creature of a similar nature, both of them having a distant resemblance to the Camel. Its body is covered with very fine long wool which is much valued; its colour that of dried roses or a dull purple; the body and feet white: they live in vast herds, and inhabit the most elevated parts of the highest mountains of Peru, where they endure the utmost rigour of frost and snow.

One of these animals has lately arrived in London, and exhibits a curious and elegant example of this tribe of quadrupeds. It is lively and elegant in its form, and seems to constitute a species between the Antelope and the Camel, and to fill up the space which Nature has placed between these two animals, a considerable degree of confusion and difficulty has arisen with respect to the distinctions which separate this species from the Lama and the Acalpa, both of Peru. The figures having been generally very imperfect, in order therefore to acquire a greater accuracy in this instance, we have procured an exact drawing made from the live animal; and such an opportunity is not likely often to recur.

Two Lamas, the one male, the other female, have been procured for the French Menagerie of Bonaparte, and it is said they have preserved their health exceedingly well, having been previously seasoned in the warmer climate of Barbadoes; of these an engraving has been published, which differs very materially from the delineation of Buffon, and even all other writers upon the subject. The Acalpa seems to be an' animal distinct from both the Lama and Vicuna, and is perhaps nearly extinct, as it is said to be incapable of all domestication, and has been now entirely hunted down by the Spaniards.

The manner of taking the Vicuna is singular. The animal seems to have the same dread of small waving objects, which most of the deer kind have. The Indians tie together several cords with pieces of wool or cloth hanging from them, across the narrow passes in the mountains about three or four feet from the ground ; they then drive a herd of these animals towards them, and they are so terrified by the flutter of the rags, that they dare not go forward, but huddle together and suffer themselves to be killed in great numbers.

The Acalpa is another animal of Peru, and smaller than either the Lama or Vicuna; it appears not to have been hitherto well described, which is a strong proof how much yet remains to be cleared up, respecting the natural history of that interesting region of the globe. It is credibly reported that when the Spaniards first invaded America, they found there several curious animals, which are now either wholly anniliated by the increasing population of the more solitary districts, or by the useless cruelty of their oppressors. Amongst these were several curious kinds of Dogs and Cats, and which have been lately described in a Spanish work upon the animals of Mexico and South America, that are now supposed to be entirely extinct. This may not improbably shortly be the case with the Vicuna and Lama, their place being so amply supplied by the Horse and the Cow—animals more esteemed by all Europeans for their usefulness and docility.

[This page was blank in the original]

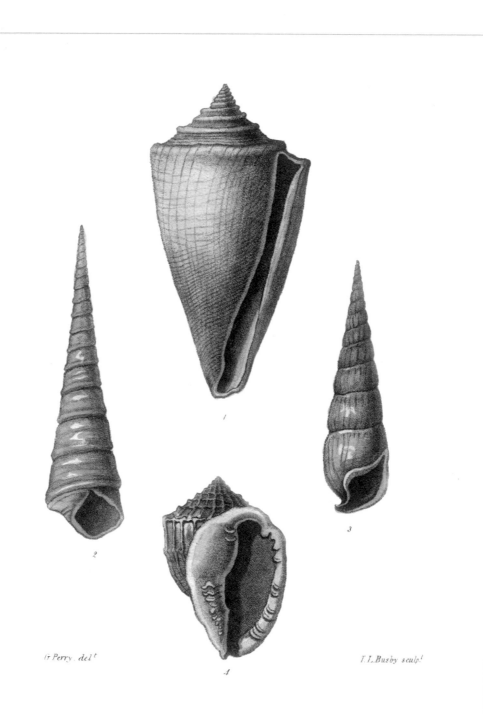

1

2

3

4

(r Perry. del.^t

T.L.Busby sculp.^t

FOSSILS.

Published by J. Stratford Holborn April 1st 1810.

PLATE 15

CLASS, FOSSILIA. ORDER, UNIVALVÆ.

No. 1. Conus angulatus, Shell univalve spiral, found in a deep bed of clay at Grignon, near Paris, also at Courtagnon in France; its form is elegant and taper, it differs considerably from any of the recent Cones at present found in the sea.

No. 2. Aculea angulata. Shell univalve spiral, the mouth, having no beak, but the base of the shell continued wholly round at the bottom.

No. 3. Cerithium lœvis. Shell univalve spiral spire, consisting of thirteen folds or revolutions.

No. 4. Cassis verrucosa. Shell oval and accuminated in the spire; the cheek or columella invested with small warty protuberances, the top decussated and reticulated; the mouth opening into a small channel.

The above shells are of the kind found in different parts of France, in beds of gravel or clay, at a considerable depth in the earth, and are in the Museum of Mr. BULLOCK in London.

REMARKS.

IT has now been concluded, that the shells and animal-remains with the Madrepores and other foreign substances which have been found in almost all the dry parts of the surface of the earth, and in a fossil state, are properly considered as monuments which bear testimony to the revolutions which the globe has undergone during the lapse of ages; and the knowledge of these fossil-remains of marine animals, and which appear to have lived in the same places where they are at present found is become highly interesting to the Naturalist.

The investigation of this subject has acquired a higher degree of importance from the light which it is calculated to throw, as well upon the *true theory of the globe*, as upon the modifications which the living Shells undergo in the regions in which they exist.

In reality it plainly appears that the fossil shells and testaceous animals, equally fossil, and the different marine fragments found in that state, scattered over very opposite regions of the globe, upon the highest mountains, in the middle of solid continents, have been the remains of animals which have resided *in these very places*, and amongst which we perceive many which have their exact similitudes, now existing in our seas.

The quantity of these animal-remains which we find in the dry parts of the globe is found to be so enormously great, that we can only suppose them to have been brought by the movement of the waters, which have formed large banks, through the extent of many strata. Amongst these remains certain shells of a tenuity and delicacy are discovered, and still retaining their form complete. These considerations seem sufficient to some writers to prove that these fossil remains *have really lived in these very parts* of the globe, and consequently that the sea has withdrawn itself from the land.

The interest with which we examine such objects increase greatly, especially when we endeavour to search the causes which has given rise to them; for we perceive plainly that the knowledge of fossils teaches us that the sea has (for a long time at least) sojourned upon parts of the globe, which are now elevated above the level of the sea, joining this to the other facts we find that they are retired now from the places formerly occupied, thus yielding to some slow but effectual cause, and from continuity of this cause, it is probable certain parts of the earth now known, will become hereafter in the lapse of time a basin for the sea, as these have already been, and so onwards, the present basin of the sea will find itself in some future age converted to a dry uncovered continent.

But we do not end here; the knowledge of fossils, by the different important facts which it presents will become the index of a perpetual change, although it is true, an infinitely slow one, which operating in all the climates, will relatively change all the surface of the globe.

Furthermore, amongst the fossil remains of animated nature found in Europe, there is evident proof, that these bodies could not have existed long in a climate such as that where they are now found.

The shells, of which the similitudes do not exist in our seas, but in the hotter climates, make up part of those found in our fossil mines. Thus the Nautilus Pompilius is found at Courtagnon and Gregnon which is one proof, and this is not the only one, the Rostellaria Fissurella is said to be found recent in the Eastern seas, (vide Martini) the Pes Pelicani also, and the Turbo Clathrus are reported (although upon a vague rumour) to be found in both the fossil and recent states, the Cypræa Sulcosa a fossil from Grignon, resembles very much the Cypræa Costata of Gmelin, vol. 5, 3413. The Oliva Triatula of Gmelin has some resemblance to the Oliva Canalifera of the Grignon fossils. The Purpura Lapillus of Courtagnon is no other than the very same shell as the Buccinum Lapillus of Linnæus, a native of the English coast. The Septa Rubecula, of Mr. PERRY's Work on Conchology, is also supposed to be the same shell which is found in a fossil state in several museums. We know also one species of Cypræa found recent on the English coast, the Cypræa Pediculus of Pennant, this is suspected to be the same which is sometimes found in a fossil state.

The above remarks are chiefly taken from the writings of Lamarck, &c. In answer to these observations of this celebrated writer, I have only to remark that his instances are *very few* which he brings of a perfect similarity existing between the fossil and recent shells. It may also be added that upon an actual examination, many of those which he seems too hastily to have judged to be similar are found to be essentially and specifically different in form. It is therefore more rational perhaps to conclude that there are no fossil shells at all similar to the recent ones, in the specific character, though there may be some general resemblance, to the eye of the spectator, who is not precisely critical in the observation of such objects.

Mr. SOWERBY has observed, very justly, that the fossil Nautilus, found at Brentford, differs very materially from the recent Nautilus Pompilius of Linnæus, and from all others yet known in a recent state; and no doubt similar distinctions exist in the others.

The Polypi, Madrepores, Marine Plants, and Oceanic Shells, attest to our senses and judgment very obviously, the revolutions which the surface of the earth has undergone in former ages. The highest mountains have undoubtedly been covered or enveloped for a certain time, by the waters of the ocean, by some great external revolution, and the sea has served no doubt as a vehicle to convey these submarine vestiges to parts, where the human mind would least expect to find such an assemblage, in beds of clay, of chalk, or limestone, and in short almost every variety of situation.

At Grignon, about seven leagues from Paris, innumerable Fossil Shells have been found in a bed of calcareous sand. At Courtagnon also near Rheims, in a sand bank of a more silicious nature than at Grignon. In Hampshire also, and the Isle of Wight, at Greenwich and Brentford numerous quantities of shells have at various times been found, inclosed in deep beds of gravel 30 or 40 feet deep.

Lamarck observes that the Fossil Shells of France and England are all of similar genera, and related to each other exactly in their form; the variety however of the French Fossils seems to be greater, several species of them have an exact resemblance in form and size, so that they seem to have been deposited by one general involution of the sea. It is very natural to suspect that they were brought from some very distant seas or coasts, and this has been indeed the general opinion of writers upon this subject, as we have no recent species in our own seas which by any means resemble them in the specific form: nevertheless many of the genera of Fossil Shells have a distant affinity to the recent kinds, at present discoverable in the Southern Seas, and which by analogy indeed may be said to prove (if such proof will be allowed) that the immense oceans of the Southern part of the World, are the depository or storehouse from whence these shells were originally driven to our northern latitudes, and there left by the retiring of the waters. These shells are scattered partially over particular

I

parts of the surface of all Europe, and the Mountains, of which they form very often a material and substantial part, are evidently of a different shape externally to all others that we are acquainted with. The arrangement of form, and the shape of their summits is always alluvial or rounded, and waved with gently undulated lines, as every soft and pulpy mass would naturally be when operated upon by a fluid, and by a successive agitation of the waters of the ocean.

This idea suggests itself to the mind upon examining the different forms of the Mountains of Portsdown, and the chalk hills which are so remarkable in the Southern countries and the internal part of England, which latter have inclined, and dipping strata of rock, and which may be supposed to possess their primæval form.

These immense Masses of chalk, according to this theory, may be considered as nothing more than the alluvial remains of some violent agitation of the waters of the ocean, and which will be more fully explained in deductions drawn from a full survey of the facts which nature presents to our view.

Remarks on the English Fossils.

IN the Fossil Reliquæ found at Willsden, Middlesex, the shells were remarkable for their form, and the fewness of the number which were found; and they were inclosed in a very solid stratum of steatite or soap rock, subsistent to bed gravel and flint stones. Several fragments of carbonaceous black wood of a fibrous texture were found with them, and several pieces of glassy *scoria and lava*, which were filled in part with air bubbles, and had evidently been at some former period in a state of fusion. The

CONCHOLOGY.

shells themselves had evidently been in a heated state from some subterraneous cause, previous to their being deposited by the great diluvial commotion which removed them from their original habitation. The inner substance of the shells seemed evidently to consist of an animal substance, incinerated and carbonized; and in some of the Tellinæ the surface was full of cracks, and discoloured in such a manner as could only happen from a sudden and violent degree of heat. These are to be considered as evident traces, in this instance, of some volcanic cause having existed at a distant period of time; and modern discoveries have confirmed the idea of volcanos being found in almost every country known at present: such is the hill of Cloud Thorpe, in Derbyshire, and of Cader Idris, in North Wales. But it is plain that volcanos are not peculiar to mountaineous countries, although any mountain that is very high and conical may be suspected to be volcanic, witness Teneriffe, the Andes and Cordilleras in Peru, with many others. In Italy, which with the surrounding islands and coasts, may be called the modern Laboratory of Volcanos; it is frequently observed even by the evidence of the present century, that many small islands and even continents, become formed from the very sea itself by the upheaving of subterraneous volcanos, causing at the same time a subsidence in some regions from the falling in of strata, and in others the accretion of mountains and hills by the gradual deposition of their mineral products. Some great external revolution of the waters must however be supposed once to have existed, to account for the phænomena of vegetable and animal remains found in continents now so far distant from the Sea. Perhaps the inclination of the earth's axis to the plane of its orbit, whether it happened *suddenly* or *gradualy*, may have caused such a change in the *direction* of the oceanic and equatorial waters of the ocean, as to make them desert some regions, leaving them capable of subsidization and vegetation for the purposes of animal and vegetable life, and invading other regions more exposed to their resistless and overwhelming force. The Flood however appears to our minds from all evidence we can derive from facts and reasoning, to have been attended with a sudden and violent impetus of the greater oceans, from some cause which afterwards gradually subsided.

If ever the axis of the Earth became suddenly altered to its present inclination by the attraction of a comet, this would at once account reasonably and theoretically for the great and awful effect. For the equatorial and southern waters which swell gradually out from the centrifugal force of the earth in its diurnal motion, would be suddenly impelled forwards, rush over and cover the whole North West continent of Europe, and certain parts of Asia and Africa. Perhaps its force might even reach over the continent of America. If the earth with a constant diurnal motion keeping this degree of inclination in its axis, went on regularly in its annual orbit, the waters would again retire from those Continents which it had invaded and enveloped, and the surface of the earth would be renewed and re-animated from those parts which had all along remained untouched and unbroken. The parts of the earth which would escape the force of such an overwhelming change of the waters, being most *distant from the Equator*, would be the whole of the large continent of New Holland, and some parts of Asia, Africa, and America, and it will be observed that no fossils are found in New Holland, nor any plants, or animals, or shells similar to any other part of the world. But in respect to those plants and animals which are found fosil in the mountains of Europe, of Asia, and of America, partially in different districts of these regions, these reliques are asserted to be *now living* in the Equatorial regions, viz. the Elephant, the Ceylon and Gangetic Crocodiles, and we suspect the Nautilus Pompilius, a remarkable shell fish (distinguished from all others by its peculiar structure) the Chætodon Equatoria, and amongst others of the vegetable tribe, the Mournful Tree (a peculiar plant well described by Jussieu) the Banana and the Bamboo. To these may be added the Triplex and Septa tribes of shells, with all the impressions of fern and submarine plants, Echini, &c. which in the living state are foreign to our northern regions, and which all writers do agree, from analogical reasoning, came from a southern and warmer climate. The Southern seas and the Equatorial regions are so extensive as to account fully for the immense quantities and varieties of fossil shells, plants, and marine exuiræ, which have at various times been discovered. Thus it will be found that the best opinion we can form of the great changes we have observed arises either from the general action of volcanoes, resulting from the chemical changes and decompositions of Nature, or from that great diluvian revolution which was owing to the flood, or perhaps from both united. Nor can we suppose these changes to be produced from any gradual subsidence of the ocean, for if so the effects would be traced in a more gradual analogy and progression, and not in such sudden and opposite extremes.

The circumstance of Fossil Shells being found in a burnt state, and which is not unfrequent in those specimens which have been found in the mountains of the Tyrol and of Hungary, are a stong proof that volcanoes did exist before that time. Nor is the probability denied by modern philosophers, that these eruptions are in general occasioned by a subterraneous communication of the sea, the latter insinuating itself through the cavities and strata of the earth, and meeting with inflammable substances. A general deluge may possibly be considered therefore as a means of at once diminishing this propensity to the formation of volcanoes, and that in the long course of future ages the number and violence of these eruptions will be gradually diminished, until nature at last shall rest in undisturbed repose. The general direction and form of the leading continents of the former earth, seems to have been much the same in their great outlines, as at the present period, all partial deviations of the coast and the efforts of volcanoes, have in general a peculiar character, operating in a very small proportion or scale, The great catastrophe which itself must be supposed to be partial, if produced by the waters of the ocean, might most probably be occasioned by the sudden inclination of the earth's axis, bringing for a certain time an equatorial flood over a great part of the globe. The objection of a modern French writer remarks that these fossil remains are so delicate in their texture and so exquisitely well preserved, that they could not have been violently hurried by the waters into their present situation, but must have alway existed there under a different climate, is unreasonable and contrary to natural appearances.

For the greatest part of these are really broken and worn away by some violent pressure, as is evident by their appearance and by the circumstances of fragments adhering often in such a way to the Fossil animal body, still preserved as could only have happened from the circumstance of the animal being flexible and alive at the time.

[This page was blank in the original]

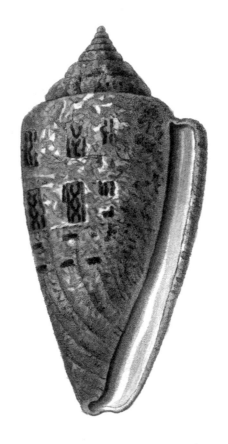

G.Perry. del.ᵗ

T.L.Busby. sculp.ᵗ

GLORIA MARIS.

Published by J. Stratford Holborn. April 1ˢᵗ 1810.

PLATE 16

CONUS, GLORIA MARIS.

Generic Character.—Shell spiral, oblong, spire short, no beak, the mouth very long and narrow, ending at the base in an open trench arcuated, the whole form of the shell cone-shaped and pointed

The shell called the Gloria Maris, on account of its dazzling beauty and the symmetry of its delicate proportions, is distinguished no less for its great rarity and the proportionably high price which it generally brings from connoisseurs. The spire consists of several rings, forming a gradual tapering summit, the body is slightly rounded, as well as the cheek of the mouth, the inside of which is white. The body is of bright olive brown, variegated with white angular spots placed irregularly; the larger marks are of a chequered pattern of a dark rich brown, of the shape of an oblong square. As these approach to the bottom of the shell, they are gently turned towards the inside, with the most pleasing mixture of all the different tint and mixed with grey, which also occurs irregularly in different parts of the exterior surface.

That superb shell which was once in the collection of the late DUCHESS of PORTLAND, was we are informed, sold to LORD TANKERVILLE, whose noble collection of shells it now adorns. Another we have observed in the British Museum, and these are, we believe, the only two of the Gloria Maris, now existing in London.

When the specimens are imperfect from age, the colours are much paler, but the peculiar shape of the spire and summit will always be a sufficiently characteristic mark to distinguish it from others of its family. By referring to the Fossil plate in our present number, the reader will perceive the similitude generally existing between the Fossil and recent shells, and how far the difference of form in these two instances, separates and sets them apart although of the same genus.

The Conus has a considerable analogy to the Genus Volutella, lately established and commonly called the Devil Shell, but the latter has a much wider mouth, and also a fluted columella, of the shape of a small screw.

Notwithstanding the great and numerous variety of species of this kind which have been described, new specimens are almost daily to be met with from the importations of the sea whalers and others, for almost every Island in the South sea seems to have its own peculiar shells, distinct from all others at present known to Conchologist.

[This page was blank in the original]

T.I. Busby. sc.   Cruikshanks. del.

KOALO.

PLATE 17

KOALO, or New Holland Sloth.

Generic Character.—Bradypus or Sloth, having five toes on each of the fore feet, and four toes on each of the hind feet; four cutting teeth in front; the body elongated, round, and covered with fine wool; the ears bushy and spreading, tipt with dark brown behind; the head flattened, round; the legs short and depressed, each foot armed with long crooked prehensile claws; the general colour cinereous, mixed with a brown tint which predominates on the back; the nose flattened and incurvated downwards: the form of the molares is unknown.

THE Bradypus or Sloth is one of those animals which are in some degree allied to the Bear, the formation of the legs and shoulders in a great measure resembling the latter. From this analogy of shape and character, the animal which has lately been discovered in the East Indies, and has been described by Bewick as the Ursine Sloth, has excited in the minds of different philosophers, an expectation of a new and more correct arrangement of their genera and species. In this hope however they have hitherto been disappointed, and we shall most probably have to wait until farther discoveries in Natural History shall enable us more accurately to define, those specimens which we at present exhibit. Even the different species of Bears are not yet thoroughly understood, those of Europe not being properly distinguished or described; but it is a point which the French writers are at present endeavouring to clear up and make more systematical.

K

ZOOLOGY.

Previous to a more particular description of the present animal, it may be necessary to observe, that although it does not agree entirely, in the form of its feet, with either the three-toed or two-toed Bradypus, which are found in other countries, yet the similitude is so strong in most peculiarities, which it possesses, that the naturalist may perhaps be considered as fully justified in placing it with the Genus Bradypus or Sloth. It is necessary to repeat, that this animal, of which there are but three or four species known, has received its name from the sluggishness and inactivity of its character, and for its remaining for a long time fixed to one spot. It inhabits woody situations, where it resides amongst the branches of trees, feeding upon the leaves and fruit, and is a solitary animal rarely to be met with. It is armed with hooked claws and the fore feet are in general longer than the hinder ones: some of the species of the Bradypus have a tail, others are without.

Amongst the numerous and curious tribes of animals, which the hitherto almost undiscovered regions of New Holland have opened to our view, the creature which we are now about to describe stands singularly pre-eminent. Whether we consider the uncouth and remarkable form of its body, which is particularly awkward and unweildy, or its strange physiognomy and manner of living, we are at a loss to imagine for what particular scale of usefulness or happiness such an animal could by the great Author of Nature possibly be destined. That the solitary and desert wastes of that immense country should be animated by creatures of so different a texture and appearance to any hitherto known, no Naturalist, however sanguine in his expectations, could have easily suspected. Many of the animals that reside in the pathless and extensive forests of New Holland, are furnished with a flap or appendage, being a winged membrane covered on the outside with hair like the rest of the body, and reaching in a square form from the toes of the fore leg to the hinder one. By the spreading out of these, they can descend, in the manner of a parachute, from branch to branch, but at the same time they have no means to fly straight forwards. Of these families are various species of Didelphis, Sciurus volans, Opossum. But it is not to be supposed that all the animals which reside amongst the branches of the trees are armed with these useful appendages of motion, for the Koalo is wholly without them and seems to have no other means than its claws, which are indeed powerful and deeply hooked for the purposes of climbing or descent.

The Koalo when fully grown is supposed to be about two feet and a half in height. [Mr. BULLOCK possesses two in his Museum, the smallest of these, it is imagined, is a young one.] The predominant colour of these animals is a bright brown or snuff colour, but suddenly growing pale towards the hinder parts or haunches. This animal, like the Capibara and some other quadrupeds, is wholly without a tail, and indeed the possession of such an appendage, in the mode of life which it enjoys, would be of little use, but rather an annoyance, as it is sufficiently defended from the flies by the length and thickness of its furry skin. The ears are dark coloured, bushy and spreading; it has four teeth projecting in front, like those of the Rabbit; but how the grinders are situated or what is their number is not hitherto known: The nose is rounded; the fore legs and underside of the belly pale and ferruginous; the eyes are sharp and sparkling; each fore foot has two thumbs and three fingers, the latter conjoined; and the hinder foot has two thumbs and two fingers, the latter conjoined; which singular combination assists them very materially in clasping hold of the branches of the trees.

ZOOLOGY.

The Kaolo is supposed to live chiefly upon berries and fruits, and like all animals not carnivorous, to be of a quiet and peaceful disposition. Its only enemies must be the Racoon and Dwarf Bear of that country, and from which it can easily escape by climbing ; and its appearance at a small distance must resemble a bunch of dry and dead moss. As there are no kind of Tygers or Wolves known as yet, except the Australasian Fox should be reckoned as a Wolf, the smaller animals must be upon the whole more secure than in most other countries.

The Koalo has more analogy to the Sloth-tribe than any other animal that has hitherto been found in New Holland, the eye is placed like that of the Sloth, very close to the mouth and nose, which gives it a clumsy awkward appearance, and void of elegance in the combination. The motions of such a creature being slow and languid, and the back lengthened out by the continual hanging posture which they assume ; they have little either in their character or appearance to interest the Naturalist or Philosopher. As Nature however provides nothing in vain, we may suppose that even these torpid, senseless creatures are wisely intended to fill up one of the great links of the chain of animated nature, and to shew forth the extensive variety of the created beings which God has, in his wisdom, constructed.

[This page was blank in the original]

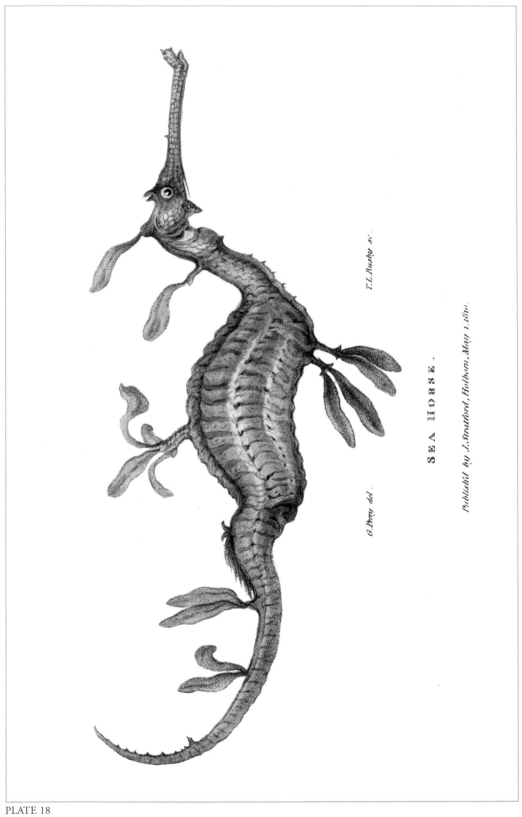

O.Perry del.

T.L.Busby sc.

SEA HORSE.

Publish'd by J.Stratford, Holborn, May 1.1810.

PLATE 18

Genus—SYNGNATHUS, or HIPPOCAMPUS.
Species—FOLIATUS.

Character.—Animal having a head formed like a Horse, the body jointed like armour, the fins placed on a pedicle irregular in their number and position, no caudal or terminating fin to the tail.

THE Hippocampus, or Sea-horse, has been always placed by the most eminent Naturalists with the Syngnathus, which last is to be considered more strictly as a fish, than the former, which is without a caudal or tail-fin. If we were to speak with more exactitude we might, not improperly, describe the Hippocampus as a marine insect, forming a distinct tribe by themselves. They have a singular resemblance in their head and neck to a Horse, and the tail may be compared in some degree to the idea which we have of a Mermaid, the nose consists of a long trunk and the mouth is small and placed at the end, the body is not covered with scales, but with a jointed kind of armour, which is divided into pentagonal plates on the back and sides, the tail is pointed at the end and divided in a similar manner. In the specimen before us the fins are shaped like leaves and are placed upon a membranaceous projecting base or prop, two and two; there is also a crest on the top of the head, and a single fin standing upon the neck; but the most remarkable fin is that which is placed on the back, as it is of a different form to the others, being oblong and placed near the commencement of the tail. This singular animal is a native of Botany Bay and is found in the seas adjacent to that curious country, it feeds in the shallow bays and coasts upon small marine insects. The Hippocampus which is found in the West Indies, differs considerably from the above in its general form, having a shorter trunk and body, and seems to swim in a more erect form. In contemplating the strange and eccentric arrangement of shapes in this singular animal, we cannot help supposing that it is possible the idea of a Dragon or Cockatrice might first have been derived from such a source, its novel and romantic outline being well calculated to impress the mind of the Painter with such an image.

The size of the Hippocampus when alive is about seven or eight inches, but there is a difference in this respect, in the male and female. The colour is of pale amber, shaded with brown, but in its living state, is said to be of a beautiful bright blue colour on the back and sides.

This circumstance is confirmed by General Davis, a gentleman whose zeal for the study and advancement of Natural History has kindly furnished us with several very useful observations.

The fish called Syngnathus, or Pipe Fish, we cannot help considering, as decidedly distinct from the proper Hippocampus, to be divided into a separate genus, and we regard the different form of the tail already described as quite a sufficient reason.

Other Naturalists, however, are of opinion, that they may both be included under the general term of Acus, our reason for differing from them will be seen in the Generic Character at the head of the Chapter.

ENTOMOLOGY.

Continuation of the History of the African Ants.

THE first object of admiration which strikes the spectator upon opening their hills, is the behaviour of the soldiers. If a breach be made in their building with a hoe or pickaxe, in a few seconds a soldier will run out and walk about the breach, as if to see wether the enemy is gone, or what is the cause of the attack. He will sometimes go in again as if to give the alarm, but most frequently in a short time is followed by two or three others, who run as fast as they can, who are soon overtaken by a large body who rush out as fast as the breach will permit them, the number increasing as long as any one batters the building. It is not easy to describe the rage and fury they shew; in their hurry they frequently miss their hold and tumble down the sides of the hill but recover themselves as quickly as possible, and bite every thing they run against. On the other hand, if they are left without interruption, they will in less than half an hour retire into the nest, as if they supposed the wonderful animal that damaged their castle was gone beyond their reach. Before the soldiers are all gone in, the labourers come forth, all in motion, and hastening towards the breach every one with a burthen of mortar in his mouth ready tempered. This they stick upon the breach as fast as they come up, and although there are thousands of them there is no hurry or confusion, but a regular wall gradually arises, filling up the chasm. Here and there a solitary soldier will be seen, who saunters about but never touches the mortar either to lift or carry it, now and then he will raise his head and with his forceps beat upon the building as if to encourage the others, upon which a loud and general hiss takes place from the labourers, who seem to hasten at such signal, redouble their pace and work as fast again. If however the assailant should renew his attack upon the building, the scene suddenly changes, a loud hiss takes place and the labourers suddenly withdraw into their pipes and galleries, in a moment they all vanish and the soldiers come forth as numerous and vindictive as before. On again ceasing with the attack they retire and the labourers once more come forth peaceably to their work. The royal chamber, where the King and Queen reside, is centrally placed and large enough to hold many hundreds of the attendants, several of these serve the purpose of nurses, for the deposting of the infant eggs which are laid by the Queen. The marching Termites are not less curious in their order than those above described, and they are larger and scarcer, they live in holes of the ground about four or five inches wide, from which they issue in vast numbers and afterwards divided into two streams or columns, twelve or fifteen abreast, and crowded like sheep, going straight forward in a direct course without deviation to the right or left. The soldiers, who are larger than the others, place themselves on each side of the path and stimulate the Ants to move forwards by a striking noise, which the army return by a loud and general hiss, and by an increased pace and motion. The Œconomy of Nature is wonderfully displayed in the species which reside under ground, which have no eyes until they arrive at a more perfect and winged state, at which time they become furnished with organs suitable to their change of situation.

The nests are formed of a dark brown clay, which when burnt affords a fine and clear red brick. Within, the whole building is pretty equally divided into innumerable cells of irregular shapes, sometimes they are quadrangular or cubic and sometimes pentagonal; but often the angles are so ill defined, that each half of the cell will be shaped like the outside of that shell which is called the Sea Ear.

Each cell has two or more entrances, and as there are galleries communicating with passages erected under ground on each side, they have in a great measure a certain place of escape to which they can retire when their principal house is destroyed. But to return to the Cities from whence these extraordinary expeditions and operations originate, it seems there is a degree of necessity for the galleries under the hills being thus large, being the great thoroughfares for all the labourers and soldiers going forth or returning from business, the fetching of clay, wood, water or provisions; and they are certainly well calculated, for the purposes to which they are applied, by the spiral slope which is given them, for if they were perpendicular the labourers would not be able to carry on the building with so much facility, as they ascend a perpendicular with great difficulty, and the soldiers can scarce do it at all. It is on this account that sometimes a road like a ledge is made on the perpendicular side of part of the building, within the hill, like those roads which are sometimes cut out of the sides of mountains, which would be otherwise inaccessible, by which and similar contrivances they travel with great facility to every part.

It has been observed before that of every species of Termites there are three orders, of these the working insects or labourers seem to be most numerous, and in the Termes Bellicosus there seems to be about one hundred labourers to one soldier or fighting insect. They are in this state about a quarter of an inch long, and from their external habits and fondness for wood have been not inexpressively called Wood Lice, by which name the French know them. They resemble them it is true very much at a distance, but they run as

L

fast or faster than any other insect of their size, and are incessantly bustling about their affairs.

The second order, or soldiers, are of a different form, having undergone a total change, they are larger than the labourers, being generally half an inch long and are supposed by some Authors to be the males. The jaws of the mouth are shaped like two very sharp awls jagged, and are capable of piercing and wounding their enemies, being as hard as a Crab's claw and placed in a strong horney head.

The third order is a winged insect, and differs from the former one in having large brown transparent wings, with which at the time of emigration it flies in search of a new settlement. In the winged state they are much enlarged in size, being now seven tenths of an inch in length. They are also furnished with two large eyes placed on each side of the head and very conspicuous, which if they have any before are not easily to be distinguished. Probably in their two first states, their eyes, if they have any, may be small, like those of the Moles, for which as they live a great part of their time under ground, they have little occasion, and are of course undistinguishable. Not only all kinds of Ants, birds and carnivorous reptiles, as well as insects, are upon the hunt for the Termites, but the inhabitants of many parts of Africa use them as food, made into a pleasant tasted pastry, with an admixture of flour.

The most remarkable circumstance in the Queen, is the great enlargement of size which takes place in the abdomen during the state of pregnancy, during which they are expanded to the length of three inches, like an oblong ball of white Cotton. This circumstance also takes place in the Pulex Penetrans of LINNÆUS, commonly called the Jigger of the West Indies, and also in the different species of Coccus or Cochineal Insec

The Termites Arborum, those which build in trees, frequently establish their nests within the roofs and other parts of houses, to which they do considerable damage if not early extirpated.

The large species are however more difficult to be guarded against, since they make their approaches chiefly under ground, descending below the foundations of houses and stores at several feet from the surface, and rising again, either in the stores or entering at the bottoms of the posts, of which the sides of the buildings are composed, boring quite through them, following the course of the fibres to the top, or making lateral cavities as they proceed.

While some are employed in gutting the posts, others ascend from them, entering a rafter, or some other part of the roof. If they once find the thatch, which is their favorite food, they soon bring up wet clay and build their pipes and galleries, through the roof in various directions as long as it will support them. In the mean time the posts will be perforated in every direction, as full of holes as that part of a ship's bottom which has been bored by worms, the fibrous and knotty parts being left to the last. The sea worms, so pernicious to shipping, appear to have the same use and office allotted them which the Termites have on land. They appear to be the most important beings in the great chain of creation, and pleasing demonstrations of that infinitely wise and gracious Power which formed the whole in harmonious order. If it was not for the rapacity of these and such other animals, tropical rivers and indeed the ocean itself would be choaked up with the bodies of trees annually carried down by rapid torrents, and as many of them would last for ages, would be productive of evils, of which we can hardly form any adequate idea.

They sometimes in carrying on their attacks, discover (although it is difficult to conceive how) that the post has some weight to support, and then if it is a convenient track to the roof, they bring their mortar and fill up all or most of the cavities, leaving the necessary roads, and as fast as they take away the wood, replace the vacancy with that material. This they work together more closely and compactly than any human art or strength could ram it, and when the house is pulled to pieces to examine the posts, the greater part is found transformed from wood to clay.

These singular insects are not less expeditious in destroying the shelves, the wainscot and other fixtures of a house than the house itself, they are particularly fond of Pine boards and Fir, which they excavate in a wonderful way, carrying away the inside and leaving only a paperlike surface, which will not weigh more than two sheets of pasteboard. On these accounts the inhabitants are careful to set their chests and boxes on stones or bricks so as to raise the bottoms above the ground, which preserves them from being so readily discovered by these insects, and also the numerous tribes of Cockroaches, Centipedes Millepedes, Scorpions and other noisome insects. Madam Merian describes a kind of Ant in the East Indies, which is smaller than the Termites, which strip the trees of their leaves, which they cut into a round form similar to a Parasol, and are seen travelling along their roads, each with one of these small coverings in his mouth, from whence they received the name of the Parasol Ants. There is also another which is found in Tobago which is highly mischievous to wooden buildings, but of which no complete description has yet been imparted by any writer upon Natural History.

[This page was blank in the original]

G.P. del.!   *T.L.B. sculp.!*

BULIMUS ZEBRA.

Published by J. Stratford, Holborn May. 1. 1810.

PLATE 19

[176]

CONCHOLOGY.

Genus—BULIMUS. *Species*—BULIMUS ZEBRA.

Character—Shell univalve, spiral, the spire and body gibbous, the summit mamillary or rounded, having no beak or rostrum, the cheek joined to the base of the columella by an undulated curve, the form of the left side of the mouth arcuated.

THE genus Bulimus has been by some writers upon Conchology placed with the Bulla or Buccinum, in the form of the spire and body, however, there is a striking difference, sufficient to distinguish it compleatly from the former. The genus Bulla has no spire protruding externally, but its revolution is involved or included internally, and the Buccinum is remarkable for a protuberant band, which is thickened and twisted upon the hinder part of the rostrum.

In the Bulimus the base of the shell is wholly joined and has no open cavity, except in front, and is therefore to be considered as wholly joined to the columella and body by a gradual rounding, forming a pleasing serpentine line. The internal surface of the mouth is grey undulated with darker shades, the outside of the shell is richly striped with purple, brown and yellow, and on the underside the body is relieved by a rich blue contrasted by gold colour.

The contrast of the colours in the specimen at present to be described, is rich and harmonious, it is moulded by the most graceful forms in Nature, a gently swelling oval predominates throughout the whole and is charmingly varied on the opposite sides of the shell. The spire is of a pale amber shade at the top and is ornamented in its differ-

ent folds with lines of a bright gold colour running round
each division.   It is a native of the South Seas and of the
Islands of New Zealand, and is much valued by collectors
for its rarity and elegant form.   The substance of the shell
is thin and bears some resemblance to that transparent appearance which is natural to the Helix, the Cyprœa, and
the denser shells, being distinguished by a superior hardness
and firmness of texture and also capable of a higher polish
upon their surface.   The Bulimus genus contains many
pleasing varieties, amongst which the minuter kinds that
have lately been discovered in the neighbourhood of New
Holland and Botany Bay, exhibit a most striking and curious tribe of shells, highly worthy of farther investigation to
the Naturalist, and in all which the analogy of the general
form is wonderfully preserved.   The above is delineated
from a specimen in Mr. BULLOCK's Museum.

The genus Melania differs from the Bulimus in having
a thick reflexed margin surrounding the whole of the mouth
and a different colour to the rest of the shell, in other
respects the form and character is very similar and not easily to be distinguished from the above.

[This page was blank in the original]

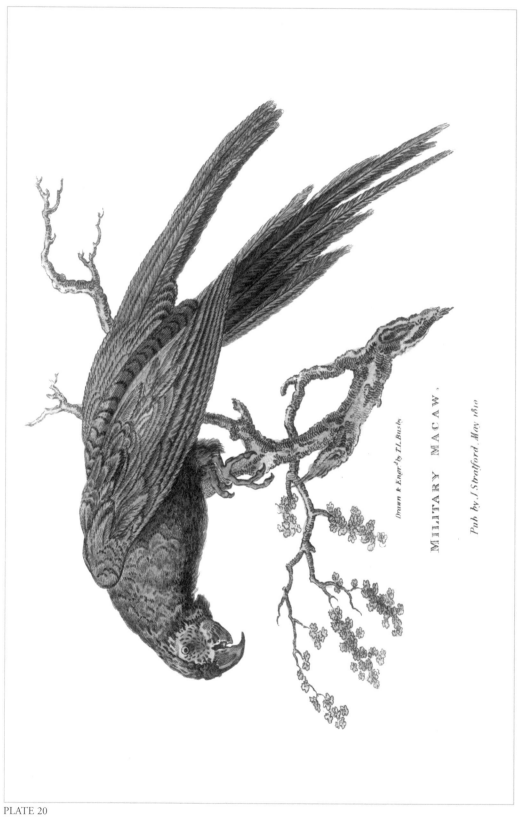

MILITARY MACAW.

Drawn & Engr.d by T.L.Bush.

Pub. by J.Stratford. May 1810.

PLATE 20

[180]

ARA MILITARIS; or MILITARY MACAW.

Character—Bill hooked, prehensile, square shaped, blunt; the under mandible closing into the upper part of the bill; the cheek covered by a circular naked membrane surrounding the eye; neck and upper coverts of an olive green, the back and other parts blue; the tail longest in the middle, cuneiform and spread out.

THE elegant and stately bird which we have selected for the present subject of observation, is of the Parrot-tribe, and is one of those singular species lately discovered in New Holland. Its form is graceful and commanding, and it has a considerable resemblance, in its general expression of character, to the majestic Eagle.

The present extended state of navigation and commerce having opened to our view, the knowledge of the most distant islands and climates, has increased in an amazing degree our numbers of the Parrot-tribe.

The circumstance of the membrance of the bill resembling that of a Vulture, and alluded to in the generic description of the Ara, leads us to admire the analogy of Nature, and at the same time a considerable agreement exists in the bills of this genus of birds, with several of the Toucan-tribes.

The Parrot seems to have been little known to the Ancients, and is only slightly mentioned by Aristotle and Onesicrites; and the Green Parroquette with a red neck, is said to have been the original bird, first brought to Europe, and found to have the curious faculty of imitating the human voice. The Parrot is distinguished by the roundness of the head and bill, from almost all other birds, also by the delicacy of his constitution, which cannot brave with safety the rigours of a Winter climate, however, by domestic attention he is enabled to endure the severity of the European Winter, and to repay by affection and sympathy, the care and regard of his keeper.

The Military Macaw is distinguished chiefly by the following colours. The crown of the head is of an olive green, the wings of the same colour, the tail of a pale blue tint shaded on the top with streaks of maroon red, on the upper bill there is a small round tuft of red feathers projecting forwards, the bill and feet are of a cinereous or Ash colour; the whole bird having much the appearance of the German military uniform, from which circumstance it not inappropriately derives its distinctive name.

[This page was blank in the original]

WOMBACH.

Pub. by J. Stratford, June, 1811.

T.L.Busby sculp

cruikshanks del.

PLATE 21

[184]

ZOOLOGY.

Genus—OPOSSUM. *Species*—WOMBACH; OR, OPOSSUM HIRSUTUM.

Character.—Five cutting teeth in front, next to which two canine teeth and eight grinders; body having a pouch for the young; the tail short, concealed under the furry skin; the fore feet having five hooked toes, the hind feet only four toes.

THE Wombach is a newly discovered animal from Botany Bay, and on many accounts highly deserving the attention of the Naturalist. He is a thick short-legged animal, rather inactive in his motions, and of the size of a turnspit dog. His figure and movements, if they do not exactly resemble those of a bear, at least strongly remind us of that creature. His length from the tail to the nose is two and a half feet; the head seven inches; the tail only half an inch; the hair is coarse and about an inch and a half long, and thickest upon the loins and rump. The colour is of a light sandy brown, varying in its shades, but darkest along the back. The head is large and flattened, and when looking at the animal's full face, seems to form nearly an equilateral triangle, any side of which is about seven inches and a half; the hair lies in regular order upon its face, as if it were combed, with its ends pointing up like radii, from the nose as its centre. The mouth has whiskers all round, as also has each cheek, and the nostrils form two distinct cavities in front, placed near to the mouth, which is short and small. The fore legs are very strong and muscular, their length to

M

the sole of the paw is five inches and a half; the hind legs are less strong and muscular than these, and their length is also five and a half inches. In size the sexes are nearly the same, but perhaps the female is to be considered as being rather the heaviest.

The Wombach seems to live generally in a loose sandy soil, burrowing in the ground and concealed under the bushes, near the foot of the hills at Port Jackson. It feeds upon grass and roots, which it scrapes up with its claws, and is of a perfectly harmless and inoffensive disposition; if, however, violently offended, or teazed, it will snap at the person who provokes it. It has shewn frequently symptoms of docility and affection to its keeper, and will beg for food sometimes, by placing one its fore feet against the knee in the manner of a lapdog. This circumstance seemed to indicate, that with kind treatment, the Wombach might soon be rendered extremely tame and friendly, and probably affectionate; but let his tutor beware of giving him provocation, at least if he should be full grown. The Wombach has also been found in Furneaux's Islands, in the South Seas, also at Van Diemen's Land; and according to the according to the account given by the natives, the Wombach of the mountains is never seen during the day, but lives retired in his hole, feeding only in the night; but that of the islands is seen to feed in all parts of the day. The country which these animals inhabit, is in general very destitute of vegetation; it is therefore probable that the grass or leaves which they eat, may by no means constitute the whole of their food; but that they may also devour some of the smaller reptiles, which would serve to strengthen the supposed analogy they have to the hog species, which are well known to be graminivorous, as well as carnivorous, having their stomachs appropriated for that sort of nourishment.

ZOOLOGY.

We shall conclude our description of this curious animal with the following account of the taking of a live one, as mentioned by Mr. BASS, in the Second Volume of COLLINS's Account of New Holland:—" The Wombach has not any claim to swiftness of foot, as most men could run it down. Its pace is hobbling or shuffling, something like the aukward gait of a bear; but it bites hard, and is furious when provoked. It was in such circumstances only that I ever heard its voice, it made a low cry, between a hissing and a whizzing, which could not be heard at the distance of more than forty yards. I chaced one of them, and with my hand placed under his belly, suddenly lifted him off the ground without hurting him, and laid him upon his back along my arm like a child. It made no noise, nor any effort to escape, not even a struggle. Its countenance was placid and undisturbed, and it seemed as contented as if it had been nursed by me from its infancy. I carried the beast upwards of a mile, and often shifted him from arm to arm, sometimes laying him upon my shoulder, all of which he took in good part; until being obliged to secure his legs while I went into the brush to cut a specimen of new wood, the creature's anger arose with the pinching of the twine; he whizzed with all his might, kicked and scratched furiously, and snapped off a piece from my jacket, with his grass-cutting teeth. Our friendship was here at an end, and the creature remained implacable all the way to the boat, ceasing to kick only when it was exhausted. Besides Furneaux's Islands, the Wombach inhabits, as has been seen, the mountains to the westward of Port Jackson; in both these places its habitation is under ground, being admirably formed for burrowing, but to what depth it descends, does not seem to be ascertained. His food is not well known, but it seems probable that he varies it, according to the situation in which he may be placed. The stomachs of such as I examined, were distended with coarse wiry grass ; and as well as others, I have seen the animal scratching among the dry ricks of sea weed thrown up upon the shores, but could never discover what he was in search of. Now the inhabitant of the mountain can have no recourse to the sea shore for his food, nor can he there find any wiry grass of the islands, but must live upon the food that circumstances present to him.

" These islands, besides the Kangaroo and Wombach, are inhabited by the Porcupine Ant-Eater; a Rat with webbed feet ; Parroquettes and small birds unknown at Port Jackson, some few of which were of beautiful plumage. Black Snakes, with venomous fangs, were numerous upon the edges of the Brush. The rocks towards the sea were covered with Fur-seals of great beauty. This species seemed to approach nearest to that named, by naturalists, the Falkland Island Seal.

" In point of animated life, Nature seems (says Mr. BASS) to have acted so oddly with this and the neighbouring islands, that if their stores were thoroughly ransacked, I doubt not but the departments of Natural History would be enlarged by more new and valuable specimens than could be acquired from any land, many times their extent."— (From a specimen in Mr. BULLOCK's Museum.)

[This page was blank in the original]

Drawn & Engrav'd by T.L.Busby.

NEW HOLLAND CRANE.

Pub.d by J.Stratford, June, 1810

PLATE 22

Genus—ARDEA ; OR, CRANE. Species—ARDEA RUBICUNDA ; OR, RED-HEADED CRANE OF NEW HOLLAND.

Character.—Bill elongated, straight and pointed ; head bare of feathers on the sides ; body oblong and oval ; neck very long ; legs tall, and the thighs imperfectly covered with feathers.

THE Crane, Heron, Stork and Curlew, form a large part of the natural division of birds, called the Waders ; or, Grallæ : to these the present undescribed bird, from Botany Bay, is closely assimilated in its form and external habits, and may be very properly referred to the Ardea ; or, Crane. Its natural food is supposed to consist of fish, and various aquatic reptiles, for which it searches, with much patient care and attention, on the banks of stagnant pools and rivers in the manner of our own English Heron. For this purpose its long and taper legs are admirably calculated, and its head being placed aloft, can view to a considerable distance amongst the reeds and long grass, those objects of which it is in pursuit.

The Ardea Rubicunda has a very considerable resemblance to a bird described by Mr. EDWARDS, in his Account of Foreign Birds, under the name of the Greater Indian Crane ; but some material differences occur in the feet and crest. In the present bird there is certainly an appearance of a web between each of the toes of the foot, which in the drawing of Mr. Edwards, does not at all appear. He also exhibits it with a tuft of black feathers projecting from the back of the head all round the neck, which in this specimen is quite different.

The small and taper form of these singular birds gives them a facility of motion, suited to their situation, and which they could not otherwise attain; how inconvenient, unsuitable and heavy would the splendid tail of the Peacock be found, if we were to suppose it changed for that of the Crane? or if he had the short legs of the Woodpecker or Dotterell, how ill suited would he be to procure his necessary food?

It is observable, that in their flight these birds always contract their long neck, into a crooked line, doubling it towards their body, in order to balance it through the air, and that the action of their wings is more slow and majestic than that of most other birds. If it should so happen that no food should offer itself in the fresh water marshes or stagnant pools where they usually resort, they flock sometimes in immense numbers to the sea shore, at which time their flesh becomes rancid and disagreeable. Nature has so provided for them, however, that they are able to endure the wants of hunger for an amazing length of time, otherwise in their long periodical journies they would be wearied and exhausted from the length of fatigue.

The Red-headed Crane, measures about five feet and a half in height, and may be considered as the tallest of the Crane-kind at present known; from wing to wing it measures six feet three inches; the general colour cinereous; the pinion, tail, and chin are black; the legs of a dark brown, and the bill of an orange colour; the back of the neck has a red carunculated skin, without feathers, and in the middle thereof a circular patch of a brown colour; the form of the body ovated and oblong; and the tail ends abruptly in a sudden recurvature. It differs entirely from the Ardea Antigone of Linnæus.—(From a specimen in Mr. Bullock's Museum.)

[This page was blank in the original]

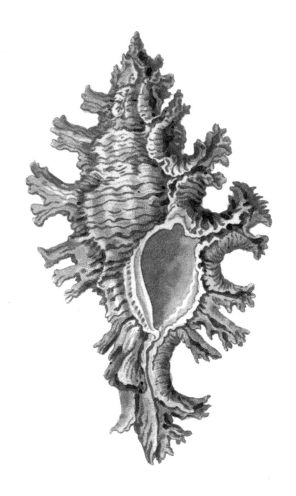

a.Perry del.   T.L.Busby sculr.

TRIPLEX FOLIATUS.

Pub by J.Stratford, June 1810.

PLATE 23

Genus—TRIPLEX. Species—TRIPLEX FOLI-ATUS.

Character.—Shell spiral univalve; the body, spire, and beak invested with three septæ or membranaceous divisions, formed into tubercles and spines; the mouth round and carunculated, colour varying from a reddish brown to a pale rose colour.

IT has been very justly remarked by different Conchologists, that the external form of Sea Shells (and indeed of the Land Shells also) affords the only certain criterion by which each genus may be distinguished. For of the shape and constitution of the animal itself we must remain for ever ignorant, as the only state in which the greatest number are obtained, is when they are empty and deserted by the animal, washed up by the force of tempests or currents of the sea. The triplex genus of Shells are remarkable for their triangulated form, which is occasioned by three thick divisions placed lengthwise on the outside of the Shell, and which form its chief ornament. Other Shells, which in many respects have a resemblance to it, are distinguished in a similar way: the Monoplex has one fold on its body; the Biplex two folds; the Hexaplex six folds, and so on with the following species, until we arrive at the greatest number, the Polyplex, in which the folds are very numerous, but the number not defined, and indeed of these latter but few have been discovered, and those only in the Southern Ocean and islands lately discovered by the investigations of Captain Cooke and other navigators.

Amongst the most agreeable and pleasing forms of Shells, which the extensive regions of the East Indies have offered to our view, we may class the Triplex Foliatus, so named from the leafy divisions and branches, forming its spines and covering its whole surface. Not indeed that there is much variety in the colours or marking of the Shell, as there are many which are more magnificently painted, but because of the elegant and taper character which it every where exhibits. The plan or structure of the Shell is three-fold, from hence its distinctive name is derived, and the folds or divisions being placed longitudinally, are spread out into branching extremities, most gracefully divided and inverted back upon the body. The mouth is embossed with a fringed edge; the rostrum or beak richly ornamented with spines of different sizes and directions, divided and pointed at the extremities. This Shell has received the common name of the Rosebush, though we cannot perceive any striking resemblance in such a comparison; the extremities of the spines are, however, often tinged with a slight shade of rose colour, which may be one cause of its receiving that name.

Several instances have occurred in specimens of this Shell, of the animal having added a fourth fold to the other three before mentioned, in which case the mouth of the Shell becomes almost closed up. This additional inclosure is very common in Shells of the Triplex kind, and is to be considered as a monstrous or unnatural accretion in the growth of the Shell, in the same manner as cows and other animals are sometimes found to have more horns than usual, and which are to be considered as deviations from the general laws of Nature.

[This page was blank in the original]

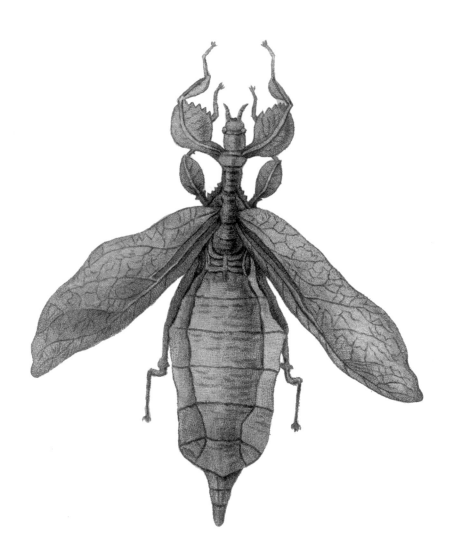

G.Perry del.ᵗ   T.L.Busby sculp.ᵗ

MANTIS FOLIATUS.

Pub.ᵈ by J.Stratford, June 1810.

PLATE 24

Genus—MANTIS. *Species*—MANTIS FOLIA-CEUS ; OR, WALKING LEAF.

Character—The antennæ filiform; the head heartshaped; six legs, the foremost with falciform hands, and a thumb of five joints; the hemelytræ folded crosswise of the length of the wings beneath them.

THE family of the Mantis differ from the insects called the Phasmata, or Spectra, in having the antennæ placed on the forehead between the eyes; whereas in the Phasmata, or Spectra, they stand on the sides of the head, far apart from each other. The legs of the Phasmata are all formed for running, like each other, and are placed so near to the head, that they are excavated near the base, to make room for the head between them. The Mantis has instead of fore legs, arms with scissar-formed hands; the upper arms and elbows are dentated or fringed.

The Mantis may also be subdivided into two families; the gouty ones, which have leaves on their legs, and the round-legged ones, which are without them. These also may be divided again into two parties; those which have round eyes, and those which have angular ones. These distinctions have been ably elucidated by Dr. LICHTENSTEIN, in the Sixth Volume of the Linnæan Transactions, and which agrees in the main with the learned FABRICIUS.

There is a remarkable difference also in the mode of life of the Phasmata and Mantis; the former live solely on

N

vegetable food, laying their eggs like Grasshoppers, in the earth; the females being furnished with a style for depositing them, of an ensiform figure, and covered by leaflets, which are found on the last division of the abdomen : the Mantis, on the contrary, confine themselves intirely to food taken from the animal kingdom ; their falciform hands serving to catch and carry to their mouths the flies and other small insects which they devour ; with regard to their metamorphosis, they never lay their eggs in the earth, but fix them on a twig, straw, or blade of grass, and these in rows or regular masses.

The insect at present figured, is from a specimen in Mr. BULLOCK's Museum, and has a striking resemblance to the form of a leaf in its wings and coverings, called the Hemelytræ. This curious circumstance, giving to the animal the appearance of a bunch of dead leaves, is undoubtedly intended for its preservation and providential escape from birds or enemies who would attack it ; its colours and form serving as a complete disguise. The providence of Nature is indeed very obvious in the same way in many other instances of the animated creation; thus the insect kingdom are found generally to be of a similar colour and appearance to the objects upon which they feed, which serves as a preventive check upon their ultimate destruction, which otherwise might too fatally ensue. Thus partly by means of defence, and partly by means of disguise and escape by flight, or other natural means, these small and seemingly insignificant creatures elude the attacks of their incroaching and formidable adversaries. The colour of the wings of these insects is infinitely varied, sometimes green, red, or brown, but always bearing a strong resemblance to the general form of a leaf; and hence they have received their characteristic name.

Extract of an Account of a Tour made at the Top of the Peake of Teneriffe, by Mr. GLAS, in the year 1761.

" The Island of Teneriffe, the highest and most conspicuous of that groupe which has been called the Canaries, is situated in the Atlantic Ocean; and the Peake, which is by much the highest point, and resembling in form a cone or sugar loaf, has been much noticed by those who have had occasion to pass near it and view its prodigious height.

" In the month of September, 1761, about four in the afternoon, I set out in company with a friend from Port Oratava, to visit the top of the Peake. We had with us a servant, a muleteer, and a guide; after ascending gradually for about six miles, we arrived at sun set at the most distant house on this side, and which stands in a hollow; here we found an aqueduct, and our servants watered the cattle, and filled some barrels with water to serve on our expedition. The valley is very beautiful, abounding with odoriferous trees and plants, and near the houses are several fields of Indian corn; and on this side of the island, the natives have two crops of grain in the year. Mounting again we travelled upon a steep road, through trees and shrubs, till we arrived at that part which is constantly overhung with clouds, and close to a large wood (it being now midnight) we alighted, made a fire, and supped; and then went to sleep under the bushes.

" The moon appearing bright, we mounted again, and travelled slowly through an excessively bad road, resembling ruins of stone buildings scattered on each side. After we got out of this part, we came upon small white light pumice-stone, like peas or shingles. Here we rode for an hour, and the wind being very cold, sharp, and piercing, our guide advised us to alight and rest till four or five in the morning.

" We followed his advice and entered into a large cave, the mouth of which was built up to about a man's height, to prevent the wind and cold from getting in. Here we found also some withered branches, with which we made a large fire to warm ourselves, and passed the time as well as we could, with one side almost scorched, and the other benumbed with cold. About five we mounted again, travelling very slowly, for the road here was very steep and rugged, till we came to a cottage built of loose stones ; the name of this place, our guide told us, was Estancia de los Inglesses (or the English Resting Place), for none but foreigners and people who gather brimstone, and by that means earn their bread, progress that far on the road. Afterwards we were obliged to alight, the road being too steep for riding, until we came to the top of a Rising or Hill, where there appeared a vast number of great loose stone, of flat surfaces, ten or more feet every way.

" Here we were compelled to travel by leaping from rock to rock, for they were not all close to each other. Amongst these is a large cavern with a well or natural reservoir, and into this we descended by a ladder, which the poor people have placed there for that purpose; part of the bottom was covered with water, but was frozen towards the inner edges ; we attempted to drink it, but could not, on account of its excessive coldness, however the guide filled a bottle which

he had purposely brought from our last station. After travelling for half a mile over the great stones or rocks, we
now were arrived at the foot of the real Peake or Sugar
Loaf, which is very steep, and to add to the difficulty of
ascending, the ground is loose and gives way under the
feet, and consequently extremely fatiguing, for although the
length of this eminence is not above half a mile, yet we
were obliged to stop and take breath I believe thirty times;
but at last we got to the top, where we lay about a quarter
of an hour to rest ourselves, being quite spent with fatigue.
The clouds were now spread out under us like an immense
ocean; and above them, at a distance, we could perceive
something black, which we took to be the Island of Ma-
deira.

" Had the air been quite clear, I have no doubt but
we could have descried Mount Atlas, in Africa, although
three hundred miles off; for though the Peak can only be
distinguished at sea at the distance of one hundred and fifty
miles, yet the spherical figure of the earth would not prevent our seeing Mount Atlas, because its summit with that
of Teneriffe, would be so far exalted above the horizon.
After we had rested for some time, we began to look about
and examine the top of the Peake, its dimensions we found
to be as Mr. EDEN describes, a hundred and forty yards in
length, and one hundred and ten in breadth. It is hollow
and shaped within like a bell subverted. From the edges
or the upper part of this bell or cauldron, as the natives call
it, to the bottom, is about forty yards. In many parts of
this hollow we observed smoke and steams of sulphur issu-
ing forth in puffs.

" The heat of the ground, in some particular places,
was so great as to penetrate through the soles of our

shoes to our feet: seeing some spots of earth or soft clay, we tried the heat with our fingers, but could not thrust them in farther than an inch or two, for the deeper we went, the more intense the heat. We then took our guide's staff and thrust it to the depth of four or five inches into a porous place where the smoke seemed to be thickest, and held it there a minute, but drawing it out we found it burnt to charcoal.

" We gathered here many pieces of most curious and beautiful brimstone of all colours, particularly azure blue, green, violet, yellow, and scarlet. The clouds had a most uncommon appearance below us, at a great distance, they seemed like the ocean, only the surface of them was not so smooth or blue, but had the appearance of very white wool. When we descended afterwards from the Peake, and entered the region of the clouds, they appeared to us as a thick mist or fog in England : all the trees of the forementioned woods, and our own cloaths were compleatly wet with it.

" The air on the top of the Peake was thin, cold, piercing, and of a dry parching nature, like the south-easterly winds which I have felt in the great Desert of Africa, or the Levanters of the Mediterranean, or even not unlike those dry easterly winds, frequent in Europe, in March or April.

" In ascending the highest part of the mountain, called the Sugar Loaf, which is very steep, our hearts panted and beat vehemently ; whether this was owing to the thinness of the air, or the uncommon fatigue of climbing, I cannot determine; perhaps it might be the two combined. Our guide, a slim agile old man, was not affected in the same manner like us, but climbed with ease like a goat ; he being one of those poor men who earn their living by gathering brimstone in the cauldron, and other volcanos ; the Peake itself being no other, although it has not burned for some years past, and all the highest parts of the island shew evident marks of those great revolutions, which have occurred in former ages.

" The Sugar Loaf itself is nothing else than earth mixed with ashes and calcareous stones, thrown out of the bowels of the earth ; and the great square stones above described, seem to have been thrown out of the cauldron or hollow of the Peake, when it was a volcano. The top is quite inaccessible on every side, except that on which we went up, which was the east. We tumbled down some large rocks towards the west, which rolled a vast way, till we lost sight of them.

" After taking some repose we began to descend, and with so much more quickness from the great descent, so that in a little more than half an hour we cleared the Peake. About five o'clock we arrived at Oratava : the whole distance from the base of the Peake we compute to be fifteen English miles ; and the hight of Estancia, above the level of the sea, I estimate at four miles ; if to this we add one mile more for the Peake, the whole height may be computed at five English miles perpendicular.

" The situation of Estancia is well adapted for the purposes of an observatory, if a warm commodious house was built upon it, to accommodate astronomers, while the moderate weather continues, viz. in July, August, and September.

"But he who visits the summit of this tremendous mountain, will find it necessary to wait for fine clear weather, to carry a good tent, and a plentiful supply both of water and provisions, so that he may remain at Estancia for four or five days, and visit the top of the Peake twice or thrice in the time, making his observations at his own leisure."

[This page was blank in the original]

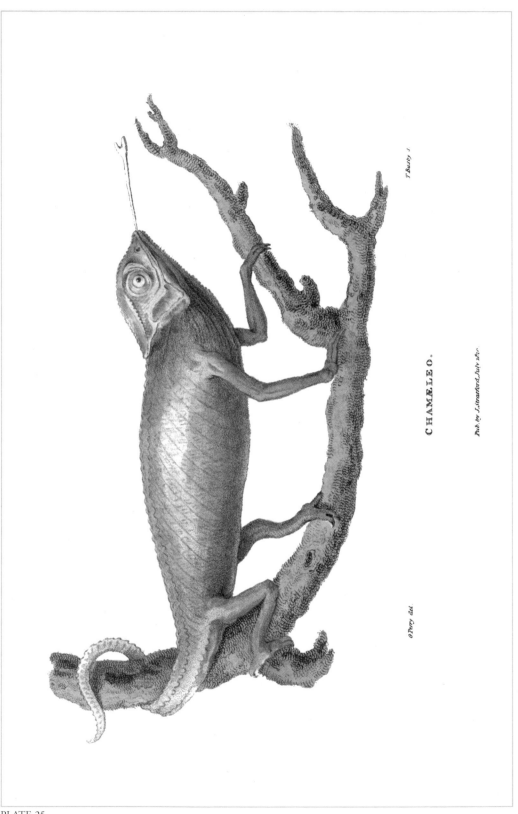

O. Perry del.

CHAMÆLEO.

Pub. by J. Stratford, July 1810.

T. Busby s.

PLATE 25

ZOOLOGY.

Genus—CHAMÆLEO. Species—CHAMÆLEO PALLIDA.

Generic Character—Body elongated, four footed, ending in a tail, head flattened, angulated, feet divided into two sections, the outer section having three toes, the inner one two, armed with short nails and prehensile.

AMONGST the various and singular shapes of Nature, which have attracted the curiosity or the wonder of travellers, none are more worthy of attention, or more apt to excite a strong interest in our minds, than the creature which we are about to describe. Its melancholy and wasted appearance indicated by its features, and the lean character of its limbs and body, would lead us naturally to consider it as one of the most miserable of created beings, which however is far fron being the case, Nature having provided, as in all other instances, for its wants and gratifications. This remarkable family of animals descri-bed under the Name of Chamæleon, has been placed by Linnæus and other Naturalists with the Lizard tribe, although certainly nothing can be essentially more different in its form, particularly in the feet and head. The feet of the Chamæleon have a considerable resemblance to those of a Parrott, and are formed to clasp the boughs of the trees, in which it chiefly resides, whereas the Lizard's foot approaches in a very great degree to the human hand, and the eyes are not capable of being elongated from the head, like those of the Chamæleon. Four or five different species have already been discovered in Africa, amongst which are the cinerea, nigra, pumila, rostrata, and the present one, the pallida from Egypt.

o

ZOOLOGY.

The general size of the Chamæleon (including the tail) is from eight inches to fifteen, in the different Species, which have been examined, and there is little doubt but that a great many more remain unknown and undiscovered in their native forests.

It is a creature quite harmless to man, and supports itself by feeding upon small insects, for which the tongue is excellently adapted, being of a contractile nature so as to be shot forth to a considerable length or drawn back into the mouth at pleasure; it is also divided at the end. By the power which it possesses, in common with the Amphibia, of inflating its chest with Air, it sometimes appears much more plump and fleshy than at other times, on this account when in its lean state its ribs may often be compleatly seen and counted as well as the Vertebræ of the back and neck: the skin is granulated, and composed of small tubercles the size of a pin's head, of an irregular Shape. The motions of the Chamæleon are extremely slow, and when sitting on a branch, or passing from one to another, it fastens itself by curling its tail round that from which it means to move, till it has secured the other by its feet. The change of Colour which always takes place upon bringing it out of a shady place into the sunshine is very remarkable, from a bluish ash colour, it becomes rather of a yellow tinge, and spotted with an appearance of Red: on reversing its body, it becomes sometimes party-coloured, one side being grey and the other brown; so that it is impossible to ascertain what exact colour is really most natural to this truly surprising animal; or to say with the Poet;

> " no Numbers can the varying robe express,
> As each new day presents a different dress."

ZOOLOGY.

Upon the return of the English expedition from the Mediterranean, the wife of an English Soldier having brought a live Chamæleon from Alexandria, supported it for several months, by keeping it in a box lined with cotton: it shewed considerable affection for its keeper, and lived chiefly upon bread soaked in milk and mixed with sugar. It was found to undergo all the relative changes of colour, which travellers have mentioned, and which have been so often disbelieved or doubted.

It is no less a remarkable circumstance that the eye of the Chamæleon is capable of being extended from the head like that of the Snail, by which a greater extension of Vision is imparted to the creature when in pursuit of its prey, as it can in all such cases give to its eyes a different inclination, the pupil being placed at the end of a muscular pivot, and moving circularly round in all directions. Its general form and appearance have not been unsuitably characterized by a modern Poet in an excellent Fable upon Credulity and Prejudice.

> A Lizard's body, lean and long,
> A Fish's head, a Serpent's tongue;
> Its claw with triple toes disjoin'd,
> And what a length of tail behind!

From the small quantity of food which the Chamæleon is found to consume it as been idly conjectured by some persons, to live entirely upon air, forming therein an exception to all the other animated tribes of nature. Upon a minute dissection however of the stomach it has been discovered that flies and the larvæ of small insects, are its general and natural food, and that it is formed with peculiar powers from nature, for undergoing a very long continuance of abstinence.

ZOOLOGY.

Philosophers have been much puzzled to account for the different changes which take place in the shades of colour, but it is most probably attributed anatomically to the secretion, or the withdrawing, of some particular fluid which exists underneath the pores of the skin, and which the animal can regulate according to its own pleasure. Such also in the human species is the nervous sensation of blushing, occasioned by the extreme afflux of blood to the extremities, or the pallid hue which results from the sudden withdrawing of the circulating fluids from extreme terror. It is not to be supposed that the Black Chamæleon can change its hue in so strong a manner as the other species, and it is accordingly found to alter only to a brown or dark purple. The Green or Olive-Coloured Chamæleon (the Chamæleon Cinerea of various authors) seems to have the greatest powers of change, and the pale species herewith described the least of all.

Of the object which the great Author of Nature had in view in such a provision of changeable appearances, it is perhaps very difficult to judge, but it is most generally and reasonably supposed to be designed for the purposes of assisting the concealment of the Animal from its external enemies, being itself a passive creature, unprovided with any weapons of offence or defence, and far more capable, by its situation, of escape than of resistance.

[This page was blank in the original]

STROMATEUS.

G.Prey del.

T.L.Busby sc.

Pubᵈ by J.Stratford, July 1810.

PLATE 26

Genus--STROMATEUS. *Species*--STROMATEUS DEPRESSUS.

Generic Character—Body oval shaped, flat and compressed, the eyes placed one on each side, one dorsal and one abdominal fin each of them commencing at the swell of the body, one lateral fin on each side near the gills, the tail divided acutely radiated.

TO the labours of the great LINNÆUS and ARTEDI, we are indebted for the most perfect investigation of the genera of Fishes, who have proposed the fins as an admirable characteristic by which to distinguish their form and mutual analogies. We shall therefore consider the other parts of each description as subordinate to the above, and regard the number and arrangement of the fins as a perfect and natural rule for finding the genus of every kind of Fish hitherto discovered in Nature. Of the genus Stromateus which we are about to describe, only four species are mentioned by BLOCH, viz. the Fiatola, Cinereus, Argenteus, and Niger; the present one being entirely new and accurately drawn from a fine specimen in the collection of Mr. WILLSHER, of Chelsea, we have denominated the Depressus from the circumstance of the singular depression of the nose, and which is not observable in the others. This Fish when fully grown, is supposed to be five inches long, and the physiognomy of its features and character are whimsical and entertaining. The lower jaw projects a little farther than the upper, the top of the head has a plate and several spines, and which are slightly united to the back fin, the eye is large and flat and placed on each side, the tail is divided into a two-fold fan, the junction being very narrow and short between it and the body.

ICHTHYOLOGY.

Such are the leading features of the Stromateus Depressus, to which we may add, that in Nature (although in its dry state this is but imperfectly shewn) the colours are delightfully vivid and pleasing. The center of the body is of a pearly colour and resembles the Opal in its prismatic variations of tints; the back, face, and tail of yellow or amber colour, and in the place of teeth there is a rough boney process on the upper and lower jaw of the Fish.

At first sight it appears to resemble a good deal, the John Dory, a Fish caught frequently in the British Seas, and which has been often celebrated as an eminent topic of conversation with epicures; but the generic character is very different and as before said peculiarly its own. If a certain method of drying and preserving the Fish of the Southern regions of the Globe, could be adopted by Naturalists and Travellers visiting those almost unknown regions, there is no doubt that such a collection might soon be formed as would tend very much to make the various productions of the great deep seem not the least, but most numerous of the animated families of the Globe, and perhaps the one least likely ever to be completely numbered. It has been indeed attempted to number the Animals and Birds, but in the numberless myriads of Fish that are spread thro' the Atlantic, Southern, and Pacific Seas, a slight comparison and view seems to be all that Mankind can ever attain to: Nature is boundless and infinite, while the knowledge of Man takes in but a small span of the wide and extensive scale of beings existing through the different stages of Animal life.

[This page was blank in the original]

Pub^d by J Stratford July 1811.

PLATE 27

Genus—DIPUS, or JERBOA. *Species*—DIPUS MUSCOLA.

Genric Character.—Incisores or cutting teeth of an irregular number, the lower ones placed horizontally, a
wide open space standing between the incisores
and the grinders, the hind legs very long, generally three times as long as the fore legs, ears
rounded and projecting, body containing an outward pouch for the young, placed like an apron
across the body.

THE Dipus or Jerboa forms a very remarkable family
of Quadrupeds, and exists in most parts of the World; in
Asia, Africa, and America, one of its strongest distinctions
is the singular length of its hinder legs, to which we may
add its pouch or apron for carrying its young, of which it
has generally two or three at a birth.

Since the discovery of the extensive Continent of New
Holland, by Capt. COOKE, and other circumnavigators,
various species of the Jerboa have been discovered in
great variety in that curious region of the Globe. Of these
the largest is the brown Kangaroo, the grey and buff kinds
being both much smaller and of different characters. Next
to these is the curious Animal called the Kangaroo Rat,
so well described in Governor PHILLIPS's Journal of a
Voyage to New Holland. Last of all appears the present
little Animal, which is considerably larger than an English
Mouse, with all the striking characters of the Jerboa or
Kangaroo, and which has been not improperly called by
the Sailors, the Kangaroo Mouse. The number and size
of its teeth is not exactly known or hitherto examined; and
it is supposed to be a very rare Animal even in its own

Country, as only one specimen is known to be in England, and which is in Mr. Bullock's Museum. The number of the teeth however is not of so much consequence in this case, because it varies in almost all the species of the genus Dipus, the general arrangement or situation however is always the same, and agrees with the character of the genus which we have given at the head of this Chapter. The Dipus Muscola is of a placid agreeable appearance and is probably capable of being tamed, like its relation the Kangaroo, as it is used to a climate similar in a great measure to that of England, it is not wholly improbable but that it may one day become a domesticated Animal. These Creatures have also the power of sitting up or resting on their hinder feet like the Squirrel and the Hare; the flesh of which they resemble a good deal in flavour, and it is said to be by no means an unpleasant kind of food.

The different species of this genus, already discovered in New Holland, amounts already to five or six, and probably other kinds may be found when the Country is more penetrated, for being of itself larger than Europe, it is natural to suppose that a great variety of Animals must inhabit so extensive a region. We ought perhaps not to omit the remark that the two inner claws of the hinder feet of this singular Animal have a ridge or narrow hollow process running down the middle of them, which makes them look as if they were double, and in which circumstance they resemble exactly the Kangaroo and other Animals brought from Botany Bay. The general height of the Animal is about eight inches, that of the Kangaroo Rat, twenty-two; the hair of the Dipus Muscola is by far the most silky and smooth in its texture, and is of a light brown colour mixed with grey.

[This page was blank in the original]

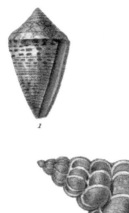

1

2

SCALARIA.

3

4

G. Perry del

T. Bushy sc.

Pub.d by J. Stratford, July 1810.

PLATE 28

Genus—SCALARIA. *Species*—SCALARIA DIS-JUNCTA.

Generic Character.—Shell spiral, univalve, the different stages of the spire rounded and separated from each other, leaving a cavity within, the spire surrounded on all sides with ribs intersectionally, placed like the steps of a ladder; the mouth round, with a flattened border. The general colour of the shell pale grey, spotted irregularly with white spots, the form altogether pointed and pyramidical.

THE extraordinary Shell which is at present a distinguished ornament of Mr. BULLOCK's Museum, is remarkable for the fine structure and shape of its folds and the curious arrangements of the parts nearest to the mouth, which seem as if quite divided from each other. The Shell is pointed, spiral, and of the form of a pyramid, environed or surrounded with transverse bands of a beautiful transparent texture; indeed the whole Shell is so thin and delicate in its formation that it is almost a miracle to find one quite perfect and unbroken in all its parts.

It is commonly named the Wentletrap, but from what derivation that word was taken, is not exactly known; the learned Author of the Enlargement of Linnæus's Systema Conchyliorum, Monsieur LAMARK, has given it the generic name of Scalaria, from its resemblance to a rope-ladder, and we have out of respect to his great talents and learning, adopted his name. Linnæus had formerly placed it with the common Turbo, and indeed at first sight it might pass for that kind of Shell, until its characteristic distinctions of the ribs or divisions, are more particularly pointed out.

P

CONCHOLOGY.

The Scalaria is generally two inches and three-quarters in length, and about one and a half in width, of a white and pale appearance inclining to grey, and it may be aptly compared to a spiral worm or screw, the folds of which do not at all touch each other, and that circumstance adds to its singularity and to the admiration it has excited among connoisseurs, so that the enormous price of fifty or a hundred pounds has frequently been offered for a fine specimen. If we were to speak of its beauty, we should say that it arises chiefly from its intricacy and transparency, in the same manner as is the case with a diamond or a beautiful piece of lace, which is admired more for rarity than beauty. As for its colours it has little to boast of in that respect, being, strictly speaking, of a tint which all artists have agreed to call neutral, and which it is impossible to describe, except by saying that it inclines to a grey. It is a native of Ceylon and Amboyna, and resides in the deepest seas of those distant regions.

Upon the whole we must consider the Scalaria as a wonderful piece of workmanship, and highly to be admired for its singularity and rarity; but in respect to the more attractive harmony of beautiful colouring, the rich tints of the Rainbow or Prism, the praise must be conferred upon other Shells, as the delight with which it fills the mind arises only from its singular shape.

The four small Shells which accompany the Wentle-trap are drawn from specimens lately imported from New South Wales, No. 1 and 2 are of the Conus genus and resemble the larger kind in their form; No. 3 is evidently of the Trochus kind; No. 4 Pyrula, resembling a little Pear. They are given to shew the variety of their patterns and form, and are hitherto unnamed by any Conchologist.

Extract from Mr. PATTERSON'S Travels in Africa, in the Year 1778.

" In the month of August we reached the Hart Beast River, situated in the Interior of Africa, in the Country of the Hottentots, and several hundred miles north of the Cape of Good Hope. The Country which we had passed through in coming from the Cape was very mountainous, and most of the Hills form Pyramids of large, loose, red, sandy stone. Here we found but few Plants in flower, except of the succulent kind. We were now arrived near the Camis Berg, a very high mountain, where we found a good supply of water; in the morning we directed our course to the West, and in our road passed several dangerous precipices. In the afternoon we came to a House belonging to a Dutchman, near a River called the Green River; here we stopped all night.

" The Hottentots were very civil and friendly, and brought us some milk, for which we exchanged with them tobacco and hemp leaves, which they prefer to the former. Their manner of living is plain and simple, nor did they seem to possess that savage and uncouth character which has been so generally attributed to them by travellers. They amused us for a part of the night by a spectacle of dancing, and in return we treated them with tobacco and dacka. Their music consists chiefly of flutes, which they form from the bark of trees of different sizes. In the afternoon we directed our course northwards, through a heavy sandy plain, which our cattle had much difficulty in crossing, and at night we came to the Great River, and our Horses being much fatigued, we waited till our waggon arrived. Here we found many Hippopotami or River Horses of immense size and bulk, and wounded one, which afterwards escaped by plunging into the stream.

" I made an excursion along the side of the Mountain, and found several new plants, the Mimosa, Salix, and several shrubby ones; amongst these may be reckoned the Euphorbia, the juice of which is supposed to be the strongest vegetable poison known in Africa: it resembles the creeping Cereus in its stalk, being prickly all over, having a small blossom adhering at the top of the stalks, which grow upright for fifteen feet. The Hottentots are supplied by this plant with poison for their arrows, by mixing it with the expressed juices of a Caterpillar, taken from another plant of the Rhus kind; sometimes for the purpose of destroying the wild Beasts, they throw the plant into certain fountains of water, frequented by them, which after drinking of the water, thus poisoned, they seldom get a thousand yards before they fall down and expire. This practice of poisoning the water, proves an additional danger to Travellers who are unacquainted with the circumstance; though the Natives generally use the precaution of leading the water which is to be poisoned into a small channel or drain, and covering up the principal fountain with boughs of the largest trees.

We next directed our Course easterly along the Banks of a river, where I added much to my collection of plants, which blossomed all around in the greatest profusion. We also beheld the most beautiful birds of gorgeous and opposite colours and numbers of Apes and Elephants.

On the fifteenth, whilst we were in this situation, Mr. Van Renan, one of my companions, had a very narrow escape of his life, in crossing at the Fording Place, he was attacked and pursued in the water by two Hippopotami or river horses, he had four Hottentots with him, and they had the good fortune to get upon a rock in the middle of the river, and their guns being loaded they killed one of those Animals; the other swam to the opposite shore. Mr. Van Renan, anxious to go to the north, in order to meet with the Camelopard, which he had heard abounded there, while I made excursions to the eastward in search of curious Plants.

I here found the Boshmen's Grass, the seed of which is used by the Boshmen as an excellent substitute for corn. Locusts at certain times of the year come also down in great quantities, so as to destroy most of the Plants, but in their journey they are themselves eaten by the Hottentots and are esteemed an excellent and delicious food.

The next day we killed an Animal of the Antelope tribe called the Hartbeast, the Capra Dorcas of Linnæus; the length of the body, including the tail and head, was five feet, six inches; it is of a brownish colour, and the flesh is palatable though dry. We then proceeded to the Sondag River, the face of the country at this place has a very barren appearance, and wild Dogs are found here which are larger than the Jackal and are very troublesome to flocks of sheep. The Hippopotami that are found here are very shy, and the chief Animals found are the Lion, Panther, Elephant, Rhinoceros, Buffalo, Antelope, &c. The Natives here are darker in their complexions but of better shape, than any I have before seen.

From the pith of a certain kind of Palm Tree the inhabitants make an excellent bread. In the neighbourhood of this place, we saw a herd of Elephants, as we conjectured eighty in number at the least, they seemed very quiet although curious, and came so near us, that we could distinguish the length and thickness of their teeth. The country is here well watered, and produces excellent pasture for Cattle. In the evening we saw a smoke upon the side of a green hill, which our guide informed us was a Caffree Village, three of the natives seemed rather alarmed at our arrival and retired to inform the whole of the villagers; they then received us kindly, brought us a present of milk, and offered us a fat Bullock, agreeably to their usual hospitable custom.

" The Village consists of fifty thatched houses and stands near a very pleasant river called Mugu Ranie and it belongs to their chief. It contains about four hundred inhabitants who are in state of vassalage to the chief, they subsist entirely on the milk of Cows and Game, not not being allowed to kill any Cattle. The men take care of the Cattle and milk them, the women cultivate all the gardens and manage the gathering of the corn.

" The Chief made us a present of a Bullock, which we afterwards shot, and the Natives were very much surprised having never seen a gun before, he wished me to have taken a hundred Bullocks in return for some Beads and Tobacco which I presented to him, he seemed half offended that I would not take them, and said, " Well, what do you think now of our Country ?" Their baskets are beautifully woven from grass by the Women, and are so close that they will hold water completely. Of these I begged for two, also two of their lances, which they freely gave me, and begged of us to stop with them for a few days, but as the weather was very hot, we chose rather to sleep in the woods, than in their Huts, and only remained there one night. They make a kind of punch, which is very pleasant, from the Guinea Corn; they make use also of Plantain, called by

Dr. Thunberg, Helaconia Caffraria. The Men are all six feet high, strong and of great courage in attacking Lions or any Beasts of prey, their skin is jet black and their clothing is made from the hides of Oxen, and they are very fond of tame Dogs, which they always keep by them.

" In the Plain we observed a very large Tree of the Mimosa kind, and soon after we came up with six Camelopards, my friend Mr. Van Renan shot one of them, which proved to be a male, I preserved the Skin and the Skeleton,* the size I found to be as follows.—The height of his natural position from the hoof to the top of his horns, fourteen feet nine inches; the length of the body, about six feet. These Animals eat the fruit of the tallest Trees, such as the Mimosa and Wild Apricot.

" Their colour is in general red or a dark brown and white, some black and white, they have no fetlock to the hoof, they are not very swift, but can continue a long chase before they stop. At a distance they look like a decayed Tree, and are spotted in general with grey spots placed irregularly, chiefly on the back.

" From this place we returned to Bokke Veld, and arrived there in four days. We had heavy showers of rain accompanied with thunder and lightning, I found an evergreen Plant, upon which grows a fruit which the Peasants use as an ingredient for poisoning the Hyena. The process is very simple and consists in drying the fruit and grinding it to powder, which they rub over a piece of flesh and throw it in places which are infested with these fierce Animals. Upon eating the flesh, the Hyenas are so imme-

* These are now preserved in the Museum of Mr. JOHN HUNTER.

diately poisoned, as generally to be found at a very little distance from the place where it was laid: this fruit is conveyed through the whole Country, for this purpose.

" The next week we departed with a Team of fresh Oxen to the Berg Valley, and passed the Mountain called Hocksberg, the summit of which is generally covered with snow, in this part is the Parel and Draken Styne, a well watered and fruitful Country, extending far to the Southward, and almost the only article of these parts, used in Commerce is the Wine, which is nearly equal to that of the Cape.

[This page was blank in the original]

KING BIRD, of PARADICE.

Pub. by J. Stratford, Aug. 1810.

Drawn & Engraved by T.L.Busby.

PLATE 29

Genus—PARADISEA. Species—PARADISEA REGIA.

Generic Character—The bill covered at the base by irregular plumes; the side feathers irregular and far extended; head and legs resembling those of a Crow, and the feet formed for walking.

THE genus Paradisea seems to present a happy and striking resemblance of the gorgeous and splendid magnificence of Eastern pomp; the crowned summit and the varied and spreading plumage of the wings and tail, frequently expanded out to an immense length and space, indicate to our minds, an idea of grandeur which has no other parrallel in Nature. In praise however to its symmetry and suitable proportions of form, our admiration must be more limited, as the head is generally so small as to be out of all natural proportion, and the legs are coarsely plaited with scales of an unpleasing texture.

The Bird of Paradise, as far as observation has removed the veil from Nature, is found only in the regions of Papua which reach to a few degrees on each side of the Equator.

It is not to be wondered at therefore, considering its rarity, that amongst the various and delightful assemblage of Birds which the Torrid Zone has yielded to the enquiring Naturalist, none have excited a warmer admiration amongst collectors, than the Bird of Paradise. It has been imagined by some authors of repute, to have been the Phœnix of the most ancient writers; but it is more probable, perhaps,

Q

that the Greeks and Romans were wholly unacquainted with this very interesting tribe of birds.

Dr Foster has presented the learned world with a classical Dissertation on the fabulous Phœnix of Antiquity, a bird of the size of an Eagle, decorated with gold and purple plumage, and more particularly described by Pliny, as having the splendor of gold round the neck; and the rest of the body and tail consisting wholly of a prismatic mixture of different colours.

The names of these Indian birds, both in their own and the European languages, appear to attribute something of a celestial origin to them. The Portuguese navigators call them the *Passeros du Sol*, or birds of the Sun; in the same manner as the Egyptians had regarded the Phœnix as a symbol of the annual revolution of the Sun. The Inhabitants of the Island of Ternate call them *Manu co dewata*, or Birds of God; from this Indian name, the Count de Buffon has derived the modern name, *Manucode*.

The Royal, or King Bird of Paradise, is of a bright orange colour on the neck and shoulders, and is perhaps more destitute of extended feathers than any other of the species. It is also the smallest of the Paradise Birds, and usually measuring about five inches and a half in length, without reckoning the two middle feathers of the tail, which are most generally six inches long. The colour of the upper part of the back is in general scarlet; the bill of a pale yellow, and about an inch in length; its base, as well as the fore part of the head, surrounded with silken plumes; the throat and part of the breast are of a deep red, and there is a flap affixed to the side of the body, consisting of feathers elegantly fringed with white and green ends.

The quil-feathers are of a bright orange brown underneath, the tail darker and more inclining to brown. From the upper part of the rump over the middle of the tail, extend two very long naked shafts, divaricating as they extend, each terminating in a beautiful circular web; the legs are strongly formed and of a pale brown colour. This species is called the King Bird, by the Dutch; it is said not to associate much with the others, but to be of a solitary nature, feeding upon berries, particularly of the red kind, seldom or ever settling upon lofty trees, but frequenting shrubs and bushes. It is considered as a very rare bird, and is said to breed in Papua, and to emigrate thence into the small Island of Arua, or Aroo, during the dry Monsoons.

In contemplating the splendid varieties of the Birds of Paradise, their costly and magnificent decorations, which at the same time are attended by no evident utility that we can perceive to themselves, as natural creatures, or even man himself, we are led to conclude that the uses of many of them will perhaps remain for ever concealed. But considered as objects adorned externally with a small portion of their Creator's glory, and with inexpressible beauty, they may serve to shew forth to the imitative powers of man, a pleasing and powerful instance of the forcible effect of opposite colours and combinations. By such means and studies, the ornamental arts of painting and design will become gradually enlarged and improved; the painter, the sculptor, or the embroiderer, may hence adopt a variety of inventions till now unknown, and add delight to the innocent pleasures and rational existence of man.

The number of the species of these tribes of birds at present known in the East Indies, is about seventy or eighty; and the European bird which seems to have the nearest affinity to them is the Hoopoe, sometimes found in England, having an elegant crest upon the summit of its head, which it can elevate or depress at pleasure. Like many of the Birds of Paradise, also, it is migratory, and retires for certain periods, to some warmer and southern climate, more congenial to its nature and the habits of its constitution.

The side-feathers which invest the body and neck of the Bird of Paradise, are admirably adapted to preserve the natural ballance and weight of the tail, which in most of the species is excessively long; without this circumstance, their flight upwards would have been more difficult, if not quite impossible. The chief advantage of their elasticity and taper form is to give a buoancy to the flight, and enable them to change their course more rapidly and effectually. Travellers who have seen them in their native regions, report that they have the power of shooting up suddenly to an astonishing height in the air, so as to leave all the lower tenants of the grove at a respectful distance, while the brightest beams of the sun reflected on their amber plumage, seem like a sudden flash of lightning or a transient gleam of light.

From a singularly fine specimen in Mr. BULLOCK's museum.

[This page was blank in the original]

G Perry del.   T.L.Busby sc.

VOLUTA PACIFICA.

Pub.d by J.Stratford, Aug.t 1810.

PLATE 30

[238]

Genus-VOLUTA. Species-VOLUTA PACIFICA.

Generic Character.—Shell univalve, spiral, columella, having four flutes, the apix mamillary.

AMONGST the univalve shells, which have recently been discovered in the Southern Ocean, the Voluta, which we are at present about to describe, stands eminently conspicuous. The genus Voluta is distinguished by the following striking characters; the shell is spiral and turrited; the mouth open and spread out, ending in a wide channel at the base; the columella or central pillar is invested with four flutes or bands; and the top or apex of the shell is mammillary, rounded. These characters of form will be always sufficient to distinguish it from the Murex, Conus, Volutella, and Ovula, which in other respects it a good deal resembles. The Voluta Pacifica is about four inches in length, of a beautiful gold colour, and richly variegated on the sides and top by elegant waved lines irregularly placed and of a dark brown colour; the mouth is of an amber colour, and there is also a remarkable horn or knobb, placed upon the cheek, which strongly projects, and is continued afterwards upon the folds of the spire. This shell is very rare, and has frequently been sold for eight or ten guineas, when in fine order and colour. It was discovered in one of the small islands near New Zealand, by that accurate investigator of Nature, Dr. SOLANDER, when employed upon a voyage of discovery with that illustrious circumnavigator, Capt. COOKE.

This shell is to be carefully distinguished from the genus Volutella, by its having no umbilicus, and its body being smooth and without tubercles or spires It has also a distant resemblance to the Cymbium genus, by the rounded and mamillary form of the Apex, or summit, but the spire, in this case, is not involved or covered over by the involution of the cheek. An immense variety of this genus of shells have lately been discovered in the regions of New Holland, and the adjacent coasts, agreeing in the general forms, as the Voluta Nivalis, Voluta Magnifica, Voluta Aurantia, &c. with several minuter species hitherto undescribed. These investigations have fortunately tended, by exciting the surprize of naturalists, to lead to a more exact and accurate arrangement of the conchological system of the moderns. The great LINNÆUS had not either the time or opportunities to illustrate sufficiently the necessary distinctions and analogies incidental to this numerous class of animals; the scientific world may therefore consider themselves as much indebted to the reform which has been effected in this part of the science, by the modern French writers, particularly Monsieur BRUGUIERE and LAMARCK, whose system is far more perfect and complete, although by no means opposite to that of the great LINNÆUS. Indeed it was high time that the darkness which enveloped this branch of natural knowledge, should be removed, and the whole subject presented in a more clear and consistent form.

[This page was blank in the original]

PAPILIO DEMOSTHENES.

G. Perry del.

T.L. Busby sc.

Pubd by J Stratford, Aug. 1811.

PLATE 31

Genus---PAPILIO. Species---PAPILIO DEMOSTHENES.

THERE is no division of natural history which has so much engaged the curiosity and attention even of the most indifferent observers of Nature, as the Papiliones, or Butter-flies. The lively contrast of their external plumage and the splendid tints with which Nature has in every country adorned them, has constantly fixed the palm of beauty, as existing in a short lived and apparently insignificant tribe of insects. Such pursuits, however, cannot be wholly deemed unfit for the examination of the human mind, since their curious and surprising history, arising from the three great stages of their transformation, is calculated to interest and instruct the thoughts in the comparative analogies of Nature. From the first consideration of the Silk-Worm and attention to its natural instincts, arose the manufacture of silk, for different articles of clothing, giving an immense employment to a very large part of the community in every country of the globe, and which is capable of being mixed with a thousand other substances to an extent which can hardly be conceived.

The Papilio Demosthenes, herewith described, has a considerable resemblance to the Papilio Teucer of LINNÆUS, but differs from that insect as well as from the Eurilochus of CRAMER, and being found only in the Brazils, must perhaps be considered as a distinct species. The chief difference is in the form and colour of the wings. The spots which are inserted upon the under side of the wings, are extremely rich, being black and fenced with a half-moon mark of pure white; the outer wings are of a rich purple verging to a blue, edged with black, and having at each end or corner of the wing, a large spot of an amber colour, with small streaks of white circularly placed. The appearance of the under wing is divided gradually, the tints being of a pleasing grey softening into a dark brown, and afterwards into a black.

The specimen from which the present drawing was taken was lately imported from the Brazils, by Mr. HULLETT, in whose collection it now remains, and may be considered as one of the grandest Papilio's already discovered, as there is a splendor and simplicity in the forms and colours, which cannot fail to arrest the attention of the connoisseur. We may also add that the back of the wings and body are partly covered with a silky flowing hair, that is difficult to represent, but which chiefly invests the central part of the wings.

We have before observed that other species exist, which much resemble the present, but as they are brought from the East Indies, it is impossible that they can be the same as that herewith described, independently of the difference of form and colour.

[This page was blank in the original]

OPOSSUM MOUSE.

Drawn and Engrav'd by T Busby, from a rare Specimen in the Possession of Mr Bullock.

Pub^d by J. Stratford. Aug. 1801.

PLATE 32

Genus—OPOSSUM. Species—OPOSSUM FLYING MOUSE.

Character.—Not exactly known.

IN the discoveries lately prosecuted in New Holland and the southern Islands of the globe, the Opossum Flying Mouse may justly be classed as the most extraordinary and eccentric; it resembles in so many of its qualities such a number of different animals, that it is almost impossible to determine to which it is most nearly allied. In the form of its teeth, it is similar to the Jerboa; in the flaps or membranes, uniting the foremost and hinder legs, it resembles the Flying Squirrel and Flying Dragon; in the pouch, which is placed like an apron for receiving its young, it resembles the Common and the Flying Opossum, and in the tail it differs from all other animals at present known, in having a flattened fin on each side.

The Opossum Flying Mouse has a lively sharp appearance, and is one of those animals which live in the trees and forests of Botany Bay and its neighbourhood, descending when it choses from bough to bough, and sometimes to the ground itself, by means of the two broad leathery flaps that reach along each side of its body, which are covered with hair on the upper side.

The teeth are sharp and projecting, and calculated for piercing and breaking all kinds of nuts and fruits; his hair is of a pleasing dusky brown, shining, and in some parts curling agreeably over his back.

R

ZOOLOGY.

Although he cannot be called the least of the quadru-ped-tribe which inhabit the variously-peopled earth, yet he undoubtedly may be classed amongst the smallest and most favoured, perhaps, of Nature's work, for he lives in a country where Cats " with deadly instinct never prowl."

It is our intention in a future number, to present our readers with a delineation of the natural distinctions of the animal called the Flying Squirrel of New Holland, from a specimen lately imported. It will be found in its full growth to be about twice the size of the present, and will make a curious and pleasing addition to our knowledge of that most curious and interesting country, leaving the reader to his own reflections on the wonderful and extensive varieties of Nature, as yet, perhaps, only half unfolded.

The delineation of the present animal is from a specimen belonging to Mr. BULLOCK, (not at present in his museum) of the natural size; the tail has much resemblance in form to a goose quil, being flat and tapering.

Extracts from Dr. Winterbottom's Travels.

An Account of the Religion and Superstition of the Modern Africans.

THE immense continent of Africa, except that part only where Mahommadenism has been impressed upon the faith of the natives by the Arabs, lies buried in the grossest ignorance. The Africans all acknowledge a Supreme Being, the great Creator of the Universe; but they suppose him to be endowed with too much benevolence to do any harm to mankind, and therefore think it unnecessary to offer him any homage. But from dæmons or evil spirits they apprehend great danger, and they endeavour, by all possible means to deprecate their wrath by sacrifice and offerings. These dæmons are supposed to be divided into two classes, the larger kind called the Aymins, are supposed to inhabit, chiefly, the deepest recesses of the forest; and the places dedicated to these spirits are generally such as inspire the spectator with awe, or are remarkable for their strange appearance, as immensely large trees rendered venerable by old age; rocks appearing in the midst of rivers, that have something in the form either gigantic or abrupt. Before they begin to sow their plantations they sacrifice a sheep, goat, fowl, or fish to the Aymin; for were this neglected, they are persuaded that nothing would grow there. In the instance where they sacrifice to the deity of the rock, a part is left for the dæmons, and the remainder is eaten by the votaries. If they should see any of the African Ants carrying away the meat, they imagine that they are taking it for the spirits. The inferior order of spirits are called the Griffee, these are supposed to reside in the skirts of a town, and sometimes even dwell within it.

When liquor is brought in, although there is no sacrifice made, a small part is always set apart for the Griffee; and the natives when rowing in canoes, never pass any of the sacred rocks without stopping to pour out a libation to the residence of the spirit, before they would venture to put a foot upon the Island. It was formerly the custom to perform religious duties in groves planted for the purpose, or the dark recesses of a forest were appropriated to this use; and their custom seems to be followed in Africa at the present day, where under the shade of the Wild Cotton or Pullum Tree, they assemble to perform their sacrifices and other rites.

To the Yahowoos, or evil spirits, are attributed all the misfortunes and afflictions occasionable to man; death, wounds, bruizes, and all the unlucky accidents of life are therefore supposed to be reducible from their malign influence. They therefore direct their prayers and supplications to them, as alone capable of appeasing their malevolence. Near the coast of Sierra Leone, superstition seems to acquire a greater power and influence over the human mind; at Whidah the principal national worship of the country is confined to Serpents, and the King Snake, which is much worshiped there, is said to be caught wild and capable of being tamed. They are about the length and thickness of a man's arm, beautiful in appearance, being grey covered with brown and yellow spots. They are harmless and enter boldly every house, in which is meat and drink constantly ready for them, and priests appointed to serve them. The Feteesh also may be reckoned as an important minor deity, and is represented by a snake, leopard, alligator, tree, &c. Upon the Kree Coast every person has his peculiar Feteesh, which is sometimes a goat, a fowl, a fish, &c. all which he never presumes to eat.

Some dare not eat fowls which are white, others dare not eat those which are black. But the most extraordinary worship is perhaps that of the Jackall, which is reckoned amongst their divinities, notwithstanding the number of sheep and sometimes children which they carry off.

At Ningo there is a temple dedicated to them and provided every evening with food, which these ravenous beasts are of course eager to take away. The Soasoos imagine that white is a very pleasing colour to the deity, they therefore when they pray, hold a white fowl in their hand and sometimes a white sheet of paper. The ceremonies of their funeral are accompanied by the most superstitious usages, one of the late Kings, who resided near Sierra Leone, lately died at the River Hunch, whither he had been removed for his health; the body was removed to the town and placed in the Palaver House, a message was sent to the Governor to desire his company at the funeral, the body was carried to the side of the grave, and a number of questions asked from the dead person, by different persons who stooped down to the coffin for that purpose. Pa-demba, a neighbouring chief, expressed his great grief in having lost so good a father, added, " that he and all the people wished him to stay a little longer with them, but as he had thought proper to leave them, they could not help it, but he and all the people wished him well." The umbrella which belonged to the deceased was put into the coffin, because they said he liked to walk with it; the pillow also which he generally used, was put into the grave, and each of the spectators threw in a handful of earth into the grave; as soon as it was closed, the women began a dismal cry, which lasted for a considerable time, until the Europeans had left the town.

The origin of amulets or charms is lost in deep antiquity, the Jews had their Phylacteries, the Greeks their Atropara, and the Romans their Amuleta; in Europe, at the present day, the superstitious practice of wearing amulets still prevails and great faith is reposed in them, when hung round the necks of children to preserve them from diseases.

In the Bullum and Timance towns, greegrees are placed to prevent the incursion of evil spirits or witches, these consist of pieces of rag like streamers, attatched to a long pole, and it would give great offence to remove or even to touch them. Greegrees are often placed in plantations, to deter people from stealing, and a few old rags placed upon an Orange Tree, will generally, though not always, secure the fruit as effectually as if guarded by the Hesperides. This superstitious dread of witchcraft, which may properly be considered as a mental disease, like many of those which the body is subject to, appears to acquire additional vigour by being transplanted from one country to another. Accordingly we find that in the West India Islands the belief in witchcraft is the occasion of as much if not more terror to the natives of Africa, where it is known by the name of Obi, notwithstanding all the efforts made to counteract it.

According to the vulgar prejudice entertained by the lower classes in England, the blacks are said to have naturally a very deleterious poison growing under their nails, with which they frequently destroy those who offend them. This opinion may have originated from the method practised by a tribe of Indians in Guiana, who sometimes conceal under their nail part of the kernel of a nut, which they secretly mix with the drink of any one they hate, and which proves, slowly, but certainly fatal. Capt. STEDMAN relates that by merely dipping their thumb in water, which they offer as a beverage to the object of their revenge, they infuse a slow, but certain death.

There is another strange practice, which the Europeans accuse the Africans of, which, however, as there can be no real foundation for it, is wrapped up in much mystery and obscurity. It is said they cause the body of any person to swell to a prodigious size, by only blowing upon them, this is sometimes done in so secret a manner, so as not to be observed by the injured party, at other times by blowing a certain substance through a long tube across the path of the traveller. There may indeed be some foundation for the latter, as the natives of Guiana are known to blow through a tube six feet long, a kind of small splinter dart dipped in poison called worrara. It is probable if these arrows possess the poisonous properties attributed to them, that the whole story of blowing is only founded upon idle report, and surmises formed from the most superstitious conjectures.

The principal hinderance of improvement and obstruction of all civilization to the Africans, seems to arise chiefly from two causes, the one for want of a full and due communication with Europe, and the too-free intercourse with the Arabs or Mahommedans. The Negroes can have no desire to cultivate the knowledge or arts of more refined countries, until habit and experience convince them of their own inferiority, and as they can have no favorable opinion of virtue and knowledge going hand in hand in the traffick or examples which they have had before their eyes, in all European countries, they naturally remain in the same superstitious bigotted state, without conviction or the free exercise of their reason.

It is nothing uncommon to find many of the black nations upon the coast, strongly infected with the principles of Mahommedanism, which they learn from the books and practices of the Arabs, and as these persons profess some degree of sanctity and learning, their example is much more likely to draw them over than the boasted mercies of those who have triumphed too long in the reign of perfidy and injustice.

It is to be hoped a happier æra will now arrive, when a period of freedom, mercy, and discussion, will be held out to the ignorant and superstitious African, and that a communication, founded upon justice, shall enlighten all the distant regions of the Atlantic.

[This page was blank in the original]

G. Perry. del.

T.L.Busby. sc.

TORTOISE.

Pub. by J. Stratford, Sept 1810

PLATE 33

ZOOLOGY.

Genus—TESTUDO, or TORTOISE.
Species—TESTUDO PANAMA *Ching-Quaw.*

Character.—Body rounded and flattened, armed with scales
    geometrically arranged, resembling a coat of mail,
    legs very short, the head retractile, mouth armed
    with a hooked bill closing over the under man-
    dible.

THIS Tortoise which is here represented for the first
time, is drawn from a live specimen, at present in the possession of Capt. HOFFMAN, of Ealing; it has resided in
England for three years, and has preserved its health exceedingly well. It is one of the smallest of its kind, hitherto
discovered, and is a native of those countries of South
America, adjoining to the Isthmus of Panama, inhabiting
the fresh water rivers and pools of that region, which is
called Terra Firma. Its general and favourite food consists
of a small quantity of dressed meat; in cold weather and the
nights of winter, it is constantly wrapped up in cotton,
which has been deemed necessary to preserve it from the
intemperate climate of Britain.

There is no part of natural history which has been subject to more errors, as to particular descriptions, than the
genus Testudo; there seems indeed at first sight, to be a
sort of natural division between the Tortoise, which has
its five claws more distinct and lives wholly upon the land,
and the Turtle, which exists chiefly as a marine animal,
and in which the claws are fin-shaped, or more obscure in
their markings, as well as irregular in their number. This
division, however, of the Tortoise from the Turtle, is very

s

obscure, for several species exist, which resort both to the land and sea, or live on the edge of the larger rivers, whose waters are alternately salt and fresh. The present Tortoise from Panama (called by the natives of that country the Ching-quaw) is supposed to be hitherto wholly undescribed, it has a considerable resemblance at the first sight to the Testudo Literatus of Thunberg, but differs in the forms and markings of the back, and also in the number of plates forming the external circle. The head, back, and legs are of a bright orange colour, mixed in a very agreeable manner with dark circles of grey, the edges being of a bright gold colour.

The protection which Nature has kindly afforded to this animal, by the strong defence of its armour, is truly wonderful and striking, affording one of the strongest instances of previous skill and design. When retiring from its natural foes, it has the power of concealing its head, legs and tail under a shelly plated covering, which envelope both the upper and under side of its body. The tail is admirably contrived for balancing the motion of the feet, which answer for the purpose of fins, being webbed between the toes like those of a Duck. It is with much difficulty that when placed upon its back in the water, ever it can recover its natural position, and the strenuous efforts, which in this case it always makes, are truly entertaining; but at length by unequally extending its feet and a constriction of the neck to one side, it overthrows the equilibrium and restores itself to the wished-for position. Upon land this is still more difficult and even impossible, the sailors therefore when they catch them upon the beach are in the habit of turning over a great number successively, and afterwards return to carry them off: their eggs also serve as an excellent food.

ZOOLOGY.

Of the sea Turtles, the most in request is the Green Turtle, so well known to epicures, which amongst other eminent discoveries of the moderns, is now esteemed a most wholesome and delicious food. About forty sloops are employed by the inhabitants of Port Royal in Jamaica, in the fishery; and as the account of the manner of taking them is rather interesting, we shall insert it at length.

" The inhabitants of Bahama, who are very expert at the art, proceed in small boats to Cuba and the adjoining islands, where in the evening, especially on moonlight nights, they watch the return of the Turtles to and from their nests, some are so large that it takes three men to turn one of them over. At other times they strike at them with a staff or spear about twelve feet long, when tired and exhausted with the pursuit he sinks to the bottom, till those who are most expert in diving will descend and bring them to the top, while another slips a noose around their necks."

The Tortoise of Ceylon which is extremely small, but elegant in its markings, has a considerable resemblance to the Ching-quaw; or, Panama Tortoise; but cannot be considered as the same animal, being the native of so distant a country, and the description we have of it is rather imperfect.

All the land Tortoises are remarkable for their longevity and their strong retention of life, even after the head has been divided from the body, and in this respect have a striking resemblance to the Eel; some of them have been authenticated to have existed for a hundred years, and one of that age is said to be now living at the City of Oxford,

The ingenuity of man has invented from the covering of the Tortoise a great variety of pleasing and useful toys, such as snuff boxes, knife handles, combs, doors of cabinets and other articles of ornament; their only practical disadvantage, and which seems to prevent the more general use of these, is their cheapness, and their yielding in elegance of lustre to the Nacre, or Mother-of-pearl.

The ancient lyre, so much celebrated in the history of Greece and other ancient nations, derives its form from the Tortoise-shell, out of which it was originally formed by the ancient artists, and still appears in the remains of their sculpture and basso-relievos, forming a most pleasing and interesting object. The Romans also adopted the name Testudo, for one of their most celebrated military arrangements in war, which consisted in placing a phalanx of their troops, closely wedged together, in such a manner that the whole of their shields should join at the top, forming a collected covering, like that of the Tortoise, impenetrable to all the arrows, stones, or darts, with which their enemies could assail them.

Thus from obvious hints, originally suggested by the simple forms of nature, arise the grander and more complicated arrangements of man, and from these alone the arts and sciences take their source, and from the Silk-worm, the Nautilus, and the Tortoise, mankind have borrowed the most useful or celebrated inventions, improved and extended through the different ages of the world.

[This page was blank in the original]

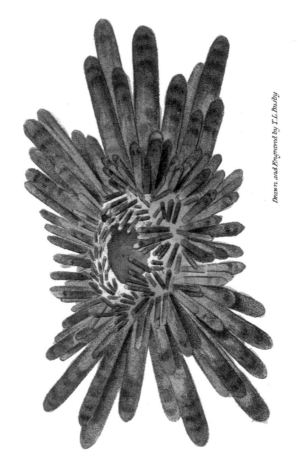

CLUB ECHINUS.

Drawn and Engraved by T.L Busby

Pub.d by J. Stratford, Sept.r 1810.

PLATE 34

Order.—NANTES; or, SWIMMERS.

Genus---ECHINUS MARINUS; or, SEA HEDGEHOG.

Species--ECHINUS CASTANEUS.

Character.—Fish inhabiting a molluscous shell or covering, circular or oval, invested with spines, placed in a radiated position and answering for the purpose of legs, by a rotary motion; an opening placed on the underside of the shell, geometrically formed, in some of the species, another opening on the top.

THE Echinus Castaneus, which we are now about to describe, belongs to a large and numerous tribe of sea animals, which bear considerable analogy to the Sea Anemone, or Animal Plant, and partly to the Polypi, in the circumstances of voluntary motion, and their radiated structure. Sir Hans Sloane, in his History of Jamaica, has described several varieties of these curious animals, which seem to unite the insect, animal, fish, and vegetable tribes. Their geometrically formed covering, attracts the eye by it's symmetry, and even when it is stripped of its spines by age or accident, it then assumes the appearance of an egg, and by its beautiful tubercles and radiations, is still interesting and delightful. In this state it is frequently found fossil, enclosed in chalk and clay, and is called the Echinus Galeatus, Echinus Cordatus, &c. and it is curious to observe that none of the fossil specimens exactly resemble the living ones, which are found at the present in the sea. This circumstance would a most lead us to suppose that there had formerly been another creation, being confirmed by many other circumstances, as that of different animals being found, fish, plants, insects, and shells, none of which have their analogies existing at present in the globe.

To return however to the more particular history of the present animal, the Echinus Castaneus, so called from Castaneus (the Chesnut,) which it exactly resembles in colour, is a native of the South Seas, and of the coasts of New Holland. It is of an oval form consisting of an arched geometrical body, ornamented with radiated spines of various lengths and of the shape of a club. These are of a flattened form, and the young or smaller ones near the center are of a purple colour. The body is small in proportion to the spines, the largest of which are about five inches long, and there is an opening at the top and bottom of the body, from which different rays issue like ribs down all the sides, having knobs or tubercles, upon which, as upon a hinge, all the spines or clubs revolve. Whether the animal has the power of moving itself by means of these spines, at the bottom of the sea, is not well ascertained, and to say the truth, they do not seem to be very well formed for such an action, though this has been the assertion of some particular travellers as well as naturalists.

The most singular animal of this tribe, is the Echinus Sceptriferus, once in the Duchess of PORTLAND's collection, and at present belonging to that curious museum of Mr. JENNINGS, of Chelsea, and which we purpose to delineate if possible in a future number; it is remarkable for having jointed spines, and is allowed to be exceedingly rare, if not quite unique, and is a native of the Eastern Seas, of Asia, and Ceylon.

[This page was blank in the original]

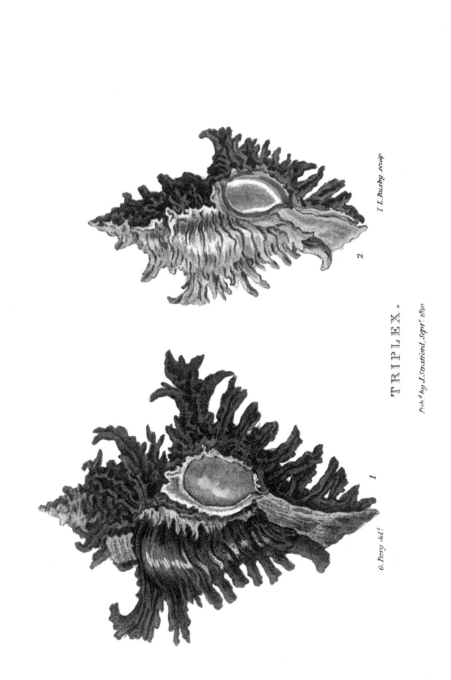

TRIPLEX.

O. Perry delt.

T. L. Busby. sculp

Pubᵈ by J. Stratford, Sepr. 1810

1

2

PLATE 35

Genus---TRIPLEX. *Species*--TRIPLEX FLAVI-CUNDA AND TRIPLEX RUBICUNDA.

Character.—Shell spiral, univalve; the body, spire, and beak invested with three septæ or membranaceous folds, formed into tubercles or spines, the mouth round and carunculated, varying in its colours in the different species.

A DELINEATION of the Triplex Foliatus has been already inserted in one of our former numbers of the ARCANA, and exhibits similar characters with the two present species. The first, which is now in the possession of Dr. COMBE, is of a singular character and colour, and has lately been discovered at Botany Bay and New Zealand. The mouth is yellow, the body dark brown, but growing paler and brighter towards the top: this shell may be considered as being very rare and valuable. The second is the Triplex Rubicunda, a scarce shell from the Island of Ceylon, the body and spire of a dark brown colour, with a bright red lip encircling the mouth, which is of a dark grey within. This shell is always rather smaller than the other, and has been improperly supposed by some collectors, as being of the same species with No. 1. although the analogy of Nature, one would suppose would sufficiently contradict such an opinion, since the red mouth Triplex is found at so considerable a distance as Ceylon and New Holland. But there is also a very considerable difference in the size and form, when examined by the eye of a critic or connoisseur.

It will perhaps be necessary to inform the reader, that the shells of the Triplex have a strong likeness or relationship to the Monoplex, Biplex, Hexaplex, Polyplex, and a number of other genera, recently elucidated and established by the Editor of this work, in a large work on the History of Shells, shortly intended to be published, and in which the generic arrangement will be upon the same improved plan.

We may perhaps be allowed to observe with what a graceful and variegated beauty Nature has adorned these elegant products of her hand, the branching forms and leafy appendages which ornament the body, the spire, and the beak, the pleasing lustre and contrast of the colours, the form of the mouth like that of a beautiful face, quite in the oval style and richly edged with pearls, which in a drawing it would be very difficult or perhaps impossible to imitate. With regard to the uses of these elegant spinous branches, we are left quite in the dark, and we may naturally suppose that they were constructed thus and thus, in order to arrest the admiration and wonder of all persons endowed with a taste for the beautiful and sublime works of art. It is a singular circumstance that the great LINNÆUS never saw more than one species of this elegant and newly discovered genus, which was the shell which he nominated the Murex Ramosus, and not being willing to make a new genus for the sake of one single shell, he crowded it amongst others, which have no natural relationship with it. It is a shell very large, nearly white all over, and is at present denominated the Triplex Ramosus.

[This page was blank in the original]

Drawn and Engraved by T.L.Busby

PAPUAN LORY.

Pub. by J. Stratford, Sept. 1810.

PLATE 36

Genus—PSITTACUS.
Species—PSITTACUS PAPUENSIS.

Generic Character.—Bill hooked and acuminated and sharply pointed, upper mandible projecting twice the length of the lower one, the head round and slightly crested, the tail long and tapering twice the length of the body; toes, standing two forward and two backward.

THE Parrot which has received the common name of the Papuan Lory, is brought from Papua in the East Indies, and is one of the richest coloured individuals of that extensive and almost numberless tribe of birds. Several naturalists having discovered that these Parrots, when in their own regions, uttered frequently a cry similar to the word lori, gave them the name of Lory, as a denomination, it is likely however shortly to go out of use, as not sufficiently distinctive in other respects, of the various birds which use that sound in particular. We have before remarked in this work how necessary it now appears to have some general reform in the historical account of the Parrots, and on this account we bestowed what we thought a due praise on Monsieur LEVAILLANT, who ably points out the tail as the most proper part for establishing a new set of orders or divisions.

The tail of the present bird has a good deal of resemblance to that of the Ara Miltaris or Military Macaw, described in the fifth number of the ARCANA. The following striking circumstances occur in the general contour of its form, distinguishing it from the rest of its congeners.

T

The head is very round, smooth, and of a highly polished appearance, and being as well as the neck, of a rich scarlet colour, is surmounted with a slight hairy crest, consisting of black and green feathers, fitting closely. The bill is pointed and sinuated, projecting over the under mandible, and of a bright chesnut colour. The shoulders are tipt with a bright spot of yellow, the middle of the back and wings are of a rich green, below this three colours occur, blue in the center, scarlet and green on each side, and the whole tail is of a dark olive colour, with the exception of the edges which are of an amber or red. In respect to elegance of form and appearance, this bird certainly claims the pre-eminence over most of his compeers, although the eye being placed very much backwards in the head gives it rather an uncouth appearance, something in the same manner as is the case with the Woodcock and Snipe. To those who are admirers of strong contrast and glaring colours, we therefore recommend the Papuan Lory, as Nature establishes varieties sometimes by elegance of form, joined to neatness of pattern; at other times, by an amazing richness and contrast of tints, but seldom bestowing therewith any very valuable qualities of voice or disposition, so that we may exclaim with one of our most favourite poets,

" Thus is Nature's vesture wrought,
To instruct the wand'ring thought."

Extracts from Barrow's Travels in China.

ON THE ARTS OF THE NATIVES.

IN respect to the arts which are invented and adopted by the Chinese, the pride and policy of the government has always been greatly inimical to the progress of the arts and sciences. The Chinese people discover no want of genius to conceive, nor of dexterity to execute, and of their imitative powers, no dispute has ever been made. Of the truth of this remark we had several instances at Yuenmin. The complicated glass lustres, consisting of several hundred pieces, were taken down, piece by piece, in the course of half an hour, by two Chinese who had never seen any thing of the kind before, and were put up again by them with similar facility, and yet it had been necessary for our mechanics to attend frequently at Mr. PARKER's warehouse, in order to be able to manage the business on their arrival in China. A Chinese undertook to cut a slip of glass from a curved piece, intended to cover the planetarium, after two of our artificers had broken three similar pieces in attempting to cut them by means of a diamond. The man performed it in private, nor could he be prevailed upon to say in what manner he accomplished it; being a little jagged along the margin, I suspect it was not cut but fractured, perhaps by passing a heated iron over a line drawn with water. It is well known that a Chinese in Canton, on being shewn an European watch, undertook and succeeded in making one like it, though he had never seen any thing like it before, but it was necessary to furnish him with a spring. The mind of a Chinese is very quick and apprehensive, and his small hands are fitted for the execution of neat work.

The manufacture of silks in China has been established from a period so remote as not to be ascertained by any history. The time, however, when cotton was first brought from the north of India into China, is noticed in their annals.

The Nankin cotton is supposed to be naturally of that colour, it having been frequently raised at the Cape of Good Hope, as an experiment, and the pods were always found to be of a buff or Nankin colour. But of all the mechanical arts, the carving of ivory has attained the greatest degree of perfection. In this branch they stand unrivalled: even at Birmingham, where I understand it has been attempted by a machine, to cut fans in imitation of the Chinese, but the experiment has not produced any articles at all equal to the other. Nothing can be more exquisitely beautiful than the fine open work displayed in a Chinese fan, the sticks of which it seems are cut singly by the hand, as a shield with the arms, or a cypher may be finished on the article at the shortest notice, and close to the drawing. From a solid ball of ivory with a hole in it not larger than half an inch across, they will cut from nine to fifteen distinct hollow globes, one within another all loosely moving, and capable of being turned round within, in all directions, and each of them carved full of open work. Models of temples, pagodas, and other pieces of architecture, are beautifully worked in ivory, in short all toys are executed in a neater manner and cheaper in China than any part of the world.

The Bamboo is useful for a thousand purposes of furniture or ornament, and the discovery of making paper from straw, although new perhaps in Europe, is of very ancient date in China, the straw of rice and other grain, the bark of the Mulberry tree, Cotton shrub, Hemp and Nettles, and other plants and materials are used in the paper manufactories in China, where they are prepared so large as to cover a whole floor. Many old persons and children earn their livelihood, by washing the ink from written paper, which being afterwards beaten and boiled to a paste, is re-manufactured into new sheets, and the ink also is saved from the water, and preserved for future use.

As to the art of Printing, there can be little doubt of its great antiquity in China, yet they have never proceeded beyond a wooden block. The nature indeed of the character is such, that moveable types would scarcely be practicable. It is true the component parts of the characters are sufficiently simple and few in number; but the difficulty of putting them together upon the frame, into the multitude of forms of which they are capable, is perhaps not to be surmounted. The power of the pulley is understood by them, but only in the single state, at least I never observed a block with more than one wheel in it. The principle of the lever should also seem to be well known, as all their valuable wares, even silver and gold are weighed with a steel-yard, and the tooth and pin iron wheels are set in motion by a water wheel. But none of the mechanical powers are applied on the great scale to facilitate or to expedite labour. Simplicity is the leading feature in their contrivances for the arts, and each tool answers several different ends. Thus the bellows of the Black-smith is nothing more than a cylinder of wood with a valvular piston, which besides blowing the fire, serves for a seat when set on an end, and as a box to contain the rest of his tools. The Barber's Bamboo basket contains his shaving apparatus, and serves when turned down as a seat for his customers. The Joiner's rule being strong, serves as a walking stick, the chest which holds his tools serves him to work on as a bench. The Pedlar's box and large umbrella serve to exhibit all his wares, and to form his little shop.

Little can be said of their art in poetry, either ancient or modern, the language being obscured so much by metaphor, as to speak rather to the eye than the ear. Of their music I have little to observe, it does not seem to be cultivated by them as a science, nor is it much cultivated by females of high life, except by those who are educated for sale, or hired out for the entertainment of others: the women generally perform on the pipe or flute, the gentlemen on guitars of two, four, or seven strings: they seem in their chorusses, to delight in the intenseness of the noise, and for this purpose the gong is admirably adapted: they have a kind of clarionet, and their kettle drums are shaped generally like barrels. The Chinese are quite unacquainted with the counter point, although they sometimes take an octave, and indeed it is not to be wondered at, as the elegant Greeks were unacquainted with it, and it was unused, even in Europe, till the monkish ages.

With regard to painting, they must be considered in two different respects. In history, as miserable daubers, unable or unwilling to execute any thing well.

In drawings of natural objects, such as flowers, birds, and insects they imitate with a great degree of exactness and brilliancy of colour, whatever is presented to their view. In landscapes they finish their pictures with great minuteness, but are deficient in those strong lights and masses of shade which give force and effect to the imitation. In the perspective delineation of buildings, there are many oversights in the arrangements of the outlines. The specimens of beautiful flowers, birds, and insects, brought over to Europe, are the work of artists at Canton, where from being in the habit of copying prints and drawings, carried thither for the purpose of being transferred to porcelain, or as articles of commerce, they have acquired a better taste than in the interior parts of the country. Great quantities of porcelain are sent from the potteries to Canton perfectly white, that the purchaser may have them of his own pattern, and specimens of these bear testimony that they are no mean copyists. In a country however, where painting is at so low an ebb, it is in vain to look for excellent works of sculpture Grotesque images of ideal beings and monstrous distortions of nature are sometimes seen upon the balustrades of bridges, and in their temples, where the niches are filled with grotesque figures of baked clay, and sometimes gilded or covered with varnish. They are as little able to model as to draw the human figure with any degree of taste or elegance; which is easily accounted for by their always drawing from themselves. Their pagodas however, have a very picturesque and pleasing effect, especially as they are generally placed on an eminence. Large four-sided blocks of stone or wood are frequently erected near the gates of cities, with inscriptions on them, meant to perpetuate the memory of certain distinguished persons. Their architecture however, in general is slight and unsolid, their pagodas being the most striking objects, the houses, and indeed the palaces of state, built very low; their temples are mostly constructed upon the same plan, with the addition of a second or a third story, standing upon the roof. The wooden pillars that constitute the colonade, are generally of Larch Fir, of no settled proportion between the length and the diameter, and they are invariably painted red and sometimes covered with a coat of varnish. Next to the pagodas is the most stupendous wall which it is supposed was raised many hundred years ago, to prevent the irruptions of the Tartars, dividing their country from all the north part of China. It is built upon the same plan as the wall of Pekin, being a mound of earth, cased on each side with bricks or stones. The astonishing magnitude of the fabric consists not so much in the plan of the work as in the immense distance of fifteen hundred miles, over which it is extended, over mountains of two or three thousand feet, in and across deep vallies and rivers. But the thick mass of the walls has been calculated, and this is found to be so great that all the materials of the houses of England and Scotland are supposed to amount to less than the bulk of the wall of China. The projecting massy towers of stone and brick are not included in this calculation. These alone are supposed to be equal to all the masonry and brick-work of London. To give another idea of the mass of matter in this stupendous fabric, it may be observed, that it is more than sufficient to surround the circumference of the earth, on two of its great circles, with two walls, each six feet high and two feet thick.

We shall now turn to a work of greater general utility, and scarce of less magnificence and grandeur. This is what has usually been called the Imperial or Grand Canal, an inland navigation of such extent and magnitude, that no other can compare therewith. The antiquity of its formation is said to be very great, it has however received many important repairs, and three of the largest rivers in the Empire carry off the superfluous water to the sea. The difficulties of such large embankments as must have been necessary for such a work, as well as the excavations, fill the mind with the greatest astonishment at the amazing perserverance and industry of the great body of the people of China.

The only parallel perhaps, which can be drawn is from the gigantic pyramids of Egypt, or in the walls of the ancient cities of Thebes or Babylon.

[This page was blank in the original]

1

2

Drawn by G. Perry   Engraved by T.L. Busby.

Pub.d by J. Stratford. Oct.r 1810.

PLATE 37

Genus---PAPILIO.   Species---PHALÆNA.

Generic Character of the Papilio—Body covered with hair, the wings plain, the antennæ armed with a capitulum or small knob placed at each extremity.

Generic Character of the Phalæna—Body covered with hair as well as the wings, the antennæ differing in the two sexes, in the male single, short, and thread-shaped, in the female long and branching out and bushy in its texture, having no knob or capitulum at the end.

IN a former number of the ARCANA, we have delineated two species of the Genus Fulgora, or Lantern Fly, and endeavoured to point out to our readers, the distinctions of its form, by which it is separated from the butterflies and moths. It is a circumstance well known to Naturalists, although not to every transient observer of the curious works of creation, that the butterfly and moth undergo several astonishing changes, previous to their acquiring their winged or perfect state. In the first place, the parent insect deposits the egg safely under some bough or leaf of a tree or shrub, which in time becomes a creeping animal called a caterpillar, this afterwards weaves for itself a warm kind of covering, in which after laying in a dormant or torpid state for several weeks or months, it bursts forth from its covering and becomes a Fly endowed with wings. This singular dormant situation is named the crysalis state, and when the creature arrives at the fly-state, it becomes a parent to a numerous progeny of eggs, which in process of time undergo the same different changes. The latter circumstances

U

of its life have been considered by moralists as a striking emblem and imitation of the soul, which after the long sleep of death, suddenly awakens with a renewed vivacity to a life of more exalted perfection and renewed existence.

To return more immediately to the subject before us, the upper fly represents the Papilio Phillis, so named by the learned insectologist Fabricius. It is a native of Mexico, and the Brazils, and is delineated from a beautiful specimen in the collection of Mr. WILLSHIRE of Chelsea. The upper wings are black with a band of red in the middle of each; there is also a yellow band, running each way from the body; upon the whole it may be considered as a very pleasing specimen of the natural family Orbati, in which all the wings are rounded in their shape.

The second represents a Moth the Phalæna Corollaria, an insect from North America, very distinguishable by its circular spots, those of the under wing being deeply shaded with blue, the general colour of the whole fly is of a soft and pleasing yellow. The Phalæna or Moth is chiefly distinguished for the soft and downy appearance of the wings and body, and in general the colours are not so gay and vivid as in the Papilio Genus.

The antennæ in this instance are branched and rounded in their outline, which circumstance characterizes the female moth, which unlike the Papilio's have their antennæ differently formed in the opposite sexes. Nothing satisfactory has hitherto been discovered of the uses of the antennæ, some authors have supposed that they are an organ of smelling, others that they are for the purpose of hearing, and it is most likely the subject will remain in doubt, until their instincts and anatomy have been in a farther degree understood.

[This page was blank in the original]

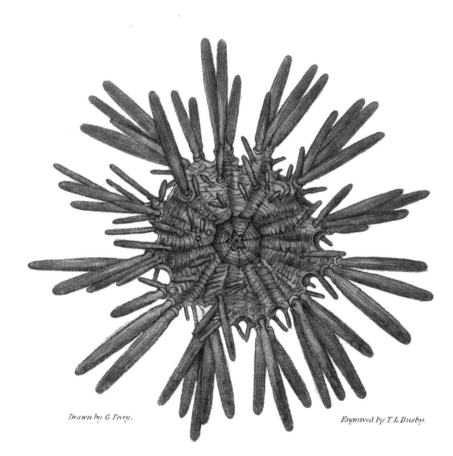

Drawn by G Perry.   Engraved by T. L. Busby.

ECHINUS STELLARIS.

Pub.d by J. Stratford, Oct. 1810.

PLATE 38

[284]

ICHTHYOLOGY.

Order--NANTES, OR SWIMMERS.
Genus--ECHINUS MARINUS, OR SEA HEDGE-
HOG. *Species*---ECHINUS STELLARIS.

Generic Character—Fish inhabiting a molluscous shell or
covering, circular or oval, invested with spines
placed in a radiated position and answering for the
purpose of legs, by a rotatory motion; an opening
placed on the underside of the shell, geometrically
formed, in some of the species another opening
on the top.

OF the various families of animated nature, which by
their natural situation are more abstruse and hid from human
investigation, may be classed the Echinus, of which we have
already described one species (the Club Echinus) in a former
number of the ARCANA. The present elegant specimen of
that curious and interesting genus is delineated from a very
perfectly preserved one brought from the South Seas, and
now deposited in Mr. Bullock's Museum. The centre of
the animal is circular and flat, having an opening at the
top and bottom, geometrically divided into six partitions,
resembling a piece of basket-work interwoven with the
young spines, shooting from the sides and the top. The
body of the animal is of an orange or flesh-colour, the lesser
spines are of a colour inclining to a pale red, and seem as if
bursting their way through the crevices of the external
surface. The largest spines are more remote from the centre,
of a dark purple brown, inclining in some of them to a grey
colour, and the opening both above and below is furnished
with an hexagonal lid, having several small openings,
through which it is supposed the animal breathes. The
form of the whole is both interesting and singular, as the

spines are placed in bunches, collectively, forming a radiated and diversified appearance, highly ornamental and pleasing. We have therefore (it being hitherto undescribed by any author) to mark it by a distinction and to separate it from its congeners, which are very numerous, given it the specific name of the Echinus Stellaris, from a fancied resemblance to the twinkling rays of a star. It may be indeed remarked as a very fortunate circumstance, that the spines should be so well preserved, as that part is the most apt to suffer from carriage and external injury. We present it therefore to our readers, undecorated by any gaudy lustre of colours, not doubting that the curious construction exhibited in its formation, will entirely atone for want of splendor. The infinite variety of the works of the Creator, as exhibited even in the les sbeautiful objects of investigation, are sufficient to excite wonder and astonishment even in the most uninformed minds, much more so with those cultivated by knowledge, who, as·the divine Shakespear has so admirably expressed it, can still find

" Books in the running brooks;
Sermons in stones, and good in every thing."

[This page was blank in the original]

Drawn by G.Perry   Engraved by T.I.Busby

CONUS PARTICOLAR.

Pub.^d by J. Stratford Octr 1810.

PLATE 39

[288]

CONCHOLOGY.

Genus-CONUS. Species-CONUS PARTICOLOR.

Generic Character—Shell univalve, spiral, oblong, spire
short, no beak, the mouth oblong and narrow,
ending at the base in an open channel rounded.
The whole form of the shell having the shape of
a cone, angular and pointed.

OF the natural history of Shell Fish, it appears that the
greatest part remains at present unknown. Of the small
microscopic shells, which do not exceed an eighth of an
inch in size, no work has hitherto been published. Of the
Fossil Shells we may also make the same observations, except
a slight essay by Dr. SOLANDER upon the Fossil Shells of
Hampshire, which can contribute little to the general
knowledge of them. Some of the very minute shells have
been published in a partial manner by Soldani Fichtel, and
Dr. Boys, but the plates are so inaccurate, that they convey
no positive information to the mind of the reader as to their
general forms.

Amongst the shells at present known to exist in the sea,
the Genus Conus affords us, perhaps the greatest numbers,
and most highly coloured varieties ; and amongst these, we
have singled out the shell at present to be described. The
Conus Particolor is a shell of a beautiful taper form resembling the Gloria Maris, formerly described in this work, in
its general shape. It is elegantly variegated with a dark
map-pattern of brown and white, and the colour of the mouth
varies in different subjects. When we contemplate the
variety and richness of colours, presented to our minds in
the tribes of Shell Fish, we cannot help suggesting who

many useful hints the ingenious artist or painter might derive from them. The forms are generally mathematical, by which circumstance the study of their shape would adapt his mind to the principles of perspective; and the harmony of the colours would unfold to his thoughts the scientific principles of Titian and Corregio: The forms of some of the Cerithiums and Terebras would serve admirably to convey to his fancy the most pleasing designs for obelisks or pyramids. Thus the study of Natural History and of the useful Arts, might be made subservient to each other, and a general taste predominate, founded only upon the true and unchangeable principles of Nature.

The idea of the Ionic capital seems to have been first derived from the examination of the fossil Cornuammonis, which is very common in Greece, and others might very probably be adopted equally ornamental and useful, in the various arts of human invention.

[This page was blank in the original]

Cruickshanks del.   P.L.Busby sc.

SANGLIN MONKEY.

Pub.d by J. Stratford, Oct. 1810.

PLATE 40

[292]

Natural Division—CERCOPITHECA, or
SANGLIN MONKEY.

Fourth Order---SAPAJUS, having prehensile Tail.
Species---SAPAJUS JACCHUS.

ONE of the most extraordinary considerations which occur to the human mind in the contemplation of nature, is the singular resemblance to the human face which appears in all the Monkey Tribes at present discovered: nor is the similitude less striking in the form of their hands, feet and bodies.

Previously to a more particular description of the present animal, it will be necessary perhaps to inform the reader, that the Monkey Tribe may not improperly be divided into four orders, according to their most striking anatomical distinctions.

First. The Simia or Ape, walking upright, having no external tail, the large toe of the foot divaricated and standing short and separate from the rest.

Secondly. The Baboons, walking oblique, having in general a short tail, the nose horizontally placed like that of a dog, in their nature fierce and untractable.

Thirdly. The Monkies distinguished by a round face and a long tail, covered with hair.

Fourthly. The Sapajus, or Prehensile Monkey, having a long tail, capable of hanging thereby to different bodies, and by coiling this round the boughs of trees or other objects, it is much assisted in its motions ; the head is variously shaped. The Sapajus, or Sapajou, is found only in the Continents of North and South America, forming a distinct race of creatures peculiar in all their habits.

Much has been discussed respecting the reasoning qualities of the Monkey ; and it has been asserted by very respectable writers in Natural History, that the Orang-Otang, when taken young, is capable of various domestic services to mankind, such as laying the table-cloth, cleaning shoes and boots, eating from a plate, &c. but these are rather to be considered as the effects of imitation than of reason, and there is no doubt that if their keepers were to provoke them to anger, that all their fancied docility and sagacity would immediately vanish, and the brutish temper quickly regain its ascendancy. The quality which is denominated Sagacity or Instinct, seems to abound much more in the Elephant, the Horse, and the Dog, approaching very nearly in these creatures to what we denominate in man, Reason.

The Sanglin Monkey, which is delineated from a live specimen in Mr. Polito's Menagerie, Exeter 'Change, is an animated and sociable little creature, not much exceeding in point of size the Squirrel-Tribe; the face is round, the nose short and flat, and his long whiskers give him a ludicrous and yet expressive physiognomy. His habits are placable, and suitable to his small powers of strength, his tail is long and narrow covered all over with short hairs. It is a native of the Brazils, and is said to subsist upon fruit, small snails and insects : like all other of its congeners its chief residence is amongst the trees, in the highest branches of the forest, where it is secured by its smallness and agility from the attacks of the larger animals of prey.

There is in the two different specimens which we have observed, a remarkable white square mark placed in the middle of the forehead, but this may perhaps not be common to the whole of the species. The animal above described used to carry on frequent battles with his keeper with nut-shells, in which he displayed singular skill and dexterity; when he was intentionally presented with a few deaf nuts, he displayed considerable contempt and anger, and conti-nued sullen for a long time after. But the most remarkable instance of an approach to reason in these creatures, is mentioned by EDWARDS in his Miscellanies. " A pair of these animals, which belonged to a merchant at Lisbon, in their state of confinement, brought forth young ones at that city. These at their birth were exceedingly curious, having no fur, they used frequently to cling fast to the teats of their dam, and when they grew a little larger they used to hang upon her back and shoulders. When she was tired she would rub them off against the wall, or whatever else was near, as the only mode of ridding herself of them. On being forced from her, the male would very affectionately take his turn of nursing, and allowed them to hang round him, for the purpose of giving ease to the dam." Monkies are excessively troublesome in the gardens and plantations of South America, as they sometimes descend in large parties from the woods, in the night, leaving the marks of their depredations too obviously in the morning.

An instance of their officious activity lately occurred at the house of a gentleman who had received a monkey of the larger kind, as a present from abroad, and who, while the family and servants were at church on a Sunday, used to amuse himself with shaking the boughs of a large apple tree, to which he was chained, over a pig-stye, and which the pigs were busily employed in devouring.

x

ZOOLOGY.

The instinct or docility of the Orang Otang exhibits however, many instances of sagacity more remarkable, and if speech had not been denied them, would have served to have filled up a more exact gradation of the human race, from man to the brute. How infinitely does the divine gift of reason elevate the faculties above the animal creation, independent of the anatomical differences, which Dr. TISSOT, the first who anatomized the Orang Otang, has so ably pointed out. The chief differences in the form of the skull are the following, the upper part is smaller and lower than in man, the brains much less in quantity, the occiputal aperture much smaller, the nose flatter, and the ears more prominent; we may also add to this that he cannot walk so erect, owing to a particular disposition of the muscles of the thighs. The Orang Otang in his wild state is a melancholy, unsocial animal, either incapable or unwilling to unite himself to those of his own race, unlike the generality of the monkey kind, a difference, which is very providentially appointed, since his strength and numbers might in that case have been obnoxious to man. As it is, he fills up that space in the chain of animated nature which gradually descends from the European to the Negro, and from the Negro to the Brute, and is calculated by his deficiencies of intellect, to raise in the mind the warmest gratitude for those wonderful attainments and advantages, which the light of reason and revelation can alone impart.

Extracts from the Travels in China, from Sir George Staunton's Account.

UPON our arrival at the town of Tacoo in the white River, a considerable guard of Chinese soldiers were destined to attend the Embassador on shore. Whenever an European went ashore from any of them, the presence of a soldier with him, announced the immediate protection of the government, and might have been intended also as a check upon his conduct. Besides the yachts intended for landing the passengers, a large quantity of river lighters were provided for the discharge of the presents. The chief conductors of the Route Chowtagin and Vantagin waited frequently upon the Embassador, not only to pay their respects to him, but to take his commands in any case requiring their accommodation or comfort. A separate table for the gentlemen in each yacht was served up in the manner and with all the delicacies of the country, and sometimes also, in an aukward imitation of English cookery. The Chinese method of dressing victuals, consisted chiefly in stews made from animal substances, divided into small square pieces mixed with vegetables, and seasoning them with a variety of savoury sauces and a combination of opposite tastes. The meat most plentiful was beef and pork. The common fowls of Europe were also common here. Among the most expensive articles and accounted the greatest delicacies at the table, were the nests of a particular kind of Swallow, which were from a very distant part of China, and the gelatinous fins of the Shark, both of which afford rich and nourishing juices, but require like the Turtle, an admixture of strong spices to be much relished. With a view to gratify, as was thought, the English appetite, instructions were given by the Mandarines to roast large pieces, such as pigs, turkies, and geese entire. This is a mode of preparing food which did not appear to have been very well executed by the Chinese cooks.

Baking bread was as little understood as roasting meat, no proper oven was to be seen in this part of the country; instead of bread, boiled rice or other grain was generally used. The rice swells considerably in boiling, and this operation is supposed to answer the purpose of fermentation of the dough in regard to bread. To each yacht were sent jars of a yellow vinous liquor, and also a distilled spirit. The management of the latter seems to be understood better than the former, for the wine was generally muddy, indifferent in taste and soon grew sour. The spirit was strong and clear, and seldom partook of any empyreumatic odour. In some of the Chinese provinces it is distilled from Millet, in others from Rice; it is called by the Chinese show-choo, which means hot wine.

Supplies were received also of peaches from the neighbourhood of Pekin, in which province they chiefly flourish; apricots, oranges, and grapes, also sugar candy and brown sugar from Fochien.

During the Embassador's stay at Tacoo, there was an interchange of visits between the Viceroy of the province and himself, he came also one hundred miles to compliment his Highness upon the occasion. He was tottering with age, but dignified, and venerable and polite without any particular restrain or parade; there was nothing particular in the present meeting, only that the tea was brought in cups with covers and infused in each cup separately, the leaves remaining at the bottom of each cup, and that the simple infusion of this herb was thought by the host, if not by the guests, preferable to its mixture with cream and sugar.

The fields in this part of the country exhibited a high state of cultivation, and were covered with the holcus sorghum, or tallest of the vegetables producing esculent grain, commonly called Barbadoes millet. It grows ten or twelve feet high, and the lowest calculation of its increase was one hundred fold.

The weather being very warm, several of the troops carrying fans with their military arms: fans are worn in China equally by both sexes and by all ranks, and this use of them at the parade, will appear less surprising to those who have observed sometimes officers in other parts of the East, exercising their battalions with umbrellas over their heads.

Amongst other instances of attention from the Viceroy, a temporary theatre was erected opposite to his Excellency's yachts. The outside was adorned with a variety of brilliant and lively colours, and an attempt by strong contrasts to encrease their effect; the inside was managed in a gay and pleasing style of ornaments; and the actors exhibited, during the day, several pantomimes and historical plays.

The Chinese have no Sunday nor even such a division as the week; the temples are however open every day for the visits of devotees. Persons of that description have from time to time made grants though to no great amount, for the maintenance of their Clergy, but none of the lands are subject to ecclesiastical tythes. A land tax to government has been substituted in the last reign, to a poll tax, as better proportioned to the faculties of individuals; a transil duty is likewise laid on goods passing from one province to another.

Near Sanchoo wheat was perceived growing for the first time by the present travellers, it was about two inches above the ground, and though on a dry sandy soil where no rain had fallen for three months, looked remarkably well. It was very neatly sown in drills or dibbled, according to the method used of late in some parts of England; that of scattering the seed by broadcast, which on very few accidental occasions only is ever practised by the Chinese, has been found by them to be attended with a considerable loss of seed as well as diminution of the crop, which when such a method is used, is apt to grow in clusters while other parts are scarcely covered. The drill method serves likewise to employ the women and children, for which little strength is required. A gentleman of the embassy calculated that the saving of the seed alone in China in this drill husbandry, which would be lost in that of broadcast, would be sufficient to maintain all the European subjects of Great Britain.

In the Province of Shan-tung were seen growing small plantations of tobacco, but more of the annual cotton plant. The cotton forms much of the cultivation of this and the adjoining southern Province of Kiang-nan nor is it much neglected in those places to the northward, where the pods can be carried to perfection before the severe frosts set in. It is not uncommon for the cultivator to lop off the tops of the cotton leaves in order to increase the number of pods and hasten their production.

In the vicinity of the River Luen and the larger lakes adjacent, we first met with the Leutze or famed fishing bird of China, which is instructed in the art or practice of supplying his owner with fish in great abundance. It is a species of the Pelican, resembling the common Cormorant. Its body is of a brown colour, with the throat white, the tail rounded and the bill yellow. On a large lake close to this river, are thousands of small boats and rafts built entirely for this species of fishery. On each boat are ten or a dozen birds, which at a signal from the owners, plunge into the water, and it is astonishing to see the enormous size of the fishes with which they return, grasped firmly in their bills. They appeared to be so well trained that it did not require either ring or cord about their necks to prevent them from swallowing any portion of their prey, except what the master was pleased to return them for encouragement and food. The boat is lightly made and carried by the shoulders of two fishermen with a pole, the birds being generally perched in the middle.

In the marshy grounds which surround the lake, we discovered the singular plant called the Nymphæa Nelumbs, a kind of large and beautiful Water Lilly. The Chinese have always held this plant in such high value, that at length they regard it as sacred. That character however has not confined it to merely useful or ornamental purposes, as it also introduced by them to the table. The seed are very numerous and like an acorn in shape, the taste more delicate than almonds. The roots are cut into slices and in the summer served with ice. They are also laid up with salt and vinegar for the winter. This plant has been supposed by some authors to be the Lotus of the Egyptians, but there are no sufficient proofs to confirm such an opinion.

The Chinese have also several other species of the Nymphæa. The flat grounds adjoining the rivers serve as plantations for the cultivation of the rice, and from these by contrivance of irrigation and flooding of all parts of the land, the Chinese husbandman will raise two crops of rice, or one of sugar, in each year, he then suffers the land to rest till the following spring, when the same process is repeated. And thus from generation to generation successive crops are reserved from the same soil, without the least idea of any necessity to let the earth lie fallow or idle for a year. The mulberry trees which are used for the cultivation of the silk worm do not seem to differ from the common mulberry trees of Europe, they are planted in rows ten or twelve feet asunder, in beds of loamy earth, thrown about a foot high above the other surface.

The insect Silk-worms are nursed in small huts erected for that purpose in the middle of the plantations, in order to be retired from all noise, for the Chinese remark that even the barking of a dog will do some injury to the worms. Some are reared however in the towns by persons who buy the leaves for that purpose, and the eggs are plac edupon paper until the period of hatching arrives.

[This page was blank in the original]

Drawn by J.C.Whichelo, from a rare Animal in the menagerie, of Mr Kendrick. Engrav'd by T.L.Busby.

Pub. by J.Stratford, Nov.1.1810.

PLATE 41

Genus---BRADYPUS.

Species--BRADYPUS STRIATUS; or, WEASEL SLOTH.

Generic Character.—Bradypus, or Sloth ; number of the claws uncertain ; four cutting teeth in front ; the body elongated, covered with wool ; ears bushy and spreading ; the head flattened in front ; the legs short and depressed ; each foot armed with long prehensile claws ; the general colour cinereous, verging to a black ; the form of the molares unknown.

THE singular forms of the Wombach and the Koalo, have already engaged our notice and attention in a former part of the present work, and which being lately discovered, have not been examined with that anatomical attention to their structure which could reasonably have been wished. The present animal bears a considerable analogy to them, and also to the Weasel in its general form, and is supposed to reside sometimes in trees, as well as upon the ground, feeding upon insects, lavæ of Moths or Caterpillars, &c. It has been exhibited alive in London, by Mr. KENDRICK, within the last few months, (from which the drawing was taken) and is reported to have been found in South America ; this fact may perhaps, however, require confirmation, as we have no doubt that at present it may be considered as a new-discovered animal, and undescribed by any author, ancient or modern. This singular creature was about two feet in length, of a dark colour of a cinereous cast of shade, black above and brown below, the centre white, with three longitudinal black stripes, the centre one thickened in the middle of the back, the eyes surrounded with a white ring, very bright and

r

sparkling, and on the centre of the forehead, a circular white spot placed in the middle between the eyes. The body covered with a long coarse bristly wool, the legs and under-side of the body of a rich pleasing brown colour.

There seems in this animal to be a general resemblance to the Ursus or Bear species, and we much regret that we had not an opportunity of a more minute examination into his form and qualities, which might have tended to throw some light upon his natural history; but by its being shortly after the present sketch was taken, removed from London, no opportunity of that kind has yet occurred.

It were much to be wished, that a Society of Naturalists were established for the express purpose of investigating and promoting all new discoveries, in respect particularly to quadrupeds, imported from distant and unexplored countries, by whom artists might be employed to draw and anatomize the various species as they occur, and offering particular rewards for the most curious kinds, or such as could be conveniently brought over the seas alive; such a subject would be highly interesting in the extension of the knowledge of physic, and would form at once a focus for the reception and dissemination of every animal hitherto unknown. A National Menagerie, if constructed upon liberal and scientific principles, would undoubtedly meet with the patronage of the public, to whom the expences of a trifling admission with money, would, in a short time, more than reimburse the incidental or annual charge of such an establishment. This would prove a permanent foundation for encreasing knowledge, and in its progress, might be found worthy of the British Nation, and an object even of Royal Patronage.

[This page was blank in the original]

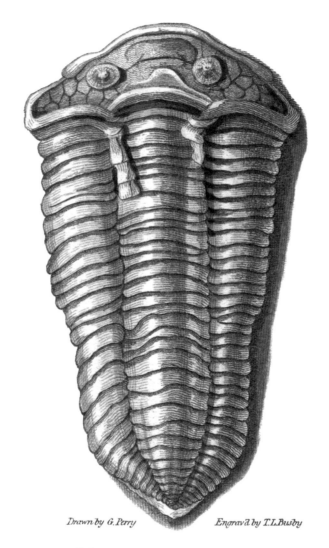

Drawn by G. Perry   Engrav'd by T.L.Busby

MONOCULITHOS.

Pub.d by J. Stratford, Nov. 1810.

PLATE 42

[308]

CONCHOLOGY.

Fossilia--FAMILY, NANTES; OR, SWIMMERS.
Genus--MONOCULITHOS. *Species*--GIGANTEA.

Generic Character.—Body tripartite, narrow at the commencement of the tail; the head gibbous and
flattened, having two projecting tubercles, resembling eyes; cheeks hexagonal and scaley; mouth
central and elongated; the body and tail covered
with arcuated ribs; the tail ending in three small
points joined firmly to a pointed base.

THIS singular petrifaction of the animal above described, supposed to have been originally marine, and deposited by the sea at the time of the Deluge, is found
plentifully in the lime-stone rocks of Dudley in Staffordshire, and other species are occasionally discovered in
Derbyshire, also in Germany and Bohemia. The difficulty
of knowing with what living creature it was most natural
to place them, has very much puzzled every naturalist,
and on this account it has been named by some, the
Monoculus Paradoxus. From a general resemblance, however, which it has to the animal called Monoculus, it has
received its present name; in which we are justified by
the remarks of Linnæus, and the late ingenious and much
lamented naturalist, Mr. Martin of Buxton, author of
an Essay on the Derbyshire Fossils.

The present curious specimen, however, is wholly
undescribed by any author whatever, being much larger
and of a different pattern to those hitherto described, for
which reason we have denominated it the Gigantea, and
Monoculithos or Monoculus turned to stone.

The specimen is of a brown colour and finely pre-served, and forms one of the most valuable and curious petrifactions of Mr. Bullock's Museum, which has in-duced us to make the drawings of three others which are smaller, in the same collection, with which it is purposed to present our readers in the ensuing Number, to shew the various species of this singular natural production.

Some light has lately been thrown upon the history of these creatures by Mr. Martin, who has discovered by microscopic inspection that the surfaces of the tubercles on the head are reticulated in their appearance, which he supposes to be a proof that they are really the natural eyes. This fact, however, we may be allowed to doubt, since in some of the individual species, six or ten, or even fifty tubercles occur upon the head, which unusual number seem contrary to the laws of nature. We should be very happy to find out that some ingenious naturalist could discover by analogy the uses and characters of the different parts and organs, at present the dissertations of these abtrusive but interesting objects, must rest in the present state till the more fortunate discovery of additional specimens, either in a living or fossil state, which may clear away the difficulties, presenting us with a series of forms, more general and connected in their analogies. In the mean time we may be allowed to express our surprize that such marine animals as these, for such they undoubtedly are, should all have become extinct by means of the flood, and that none should have been preserved in their own element.

The variety of marine insects found in various parts of the mountains of lime-stone and chalk, attest in the strongest manner the universality of the deluge. The petrified oysters, and other sea-shells so strongly prove their marine origin, that he must be of a sceptical mind who could doubt any of the circumstances. The singular species of Lizards and of Crocodiles in a fossil state, and found so abundantly in Bedfordshire, Devonshire, and Gloucestershire, would lead us to suppose that the impulse of the waters, had at the time of the deluge, brought these objects from the Southern and Pacific Ocean, and having retired again to their original beds, had left them to be covered over by the decomposition of vegetable earths. The supposition of some Philosophers who may have conceived that the axis of the earth, has at some former time, received a considerable alteration in the inclination of its line to the plane of its orbit, becomes more and more strengthened and confirmed by all the facts which the history of extraneous fossils presents to our view. The tides joined to the centrifugal force of the equatorial waters distended by the diurnal motion of the earth, may be considered as quite equal, in case of such a change, to the effect which has been produced.

In the most distant regions of Siberia and North America the immensely large bones of the Mammoth and Elephant, which have been found in a fossil state, may tend also strongly to prove some wonderful change of climate, compared with that temperature which is at present experienced by the inhabitants of the earth. If indeed the climates have become so materially altered, it accounts in some degree for the circumstances of the large fossil shells of the Cornu Ammonis kind, with its numerous varieties, being found different in form and character to any of the recent ones. The temperature of the ocean would be completely changed and become only adapted for such animals as belonged only to a colder region. The difference of organization in the fossil shell called the Orthoceras, is very obvious; this shell is very rarely found in a recent state, and then only very small, about an inch long; but in the fossil state, of an immense size, two or three feet from end to end. When curved at one end in the manner of the Pastoral Staff or Crosier, it has been called the Litnus. The Cornu Ammonis has not unfrequently been found in the Island of Portland, and in Devonshire, six or seven feet in circumference and one foot or more in thickness, embedded in different strata of quarry stone, upwards of sixty feet in depth. Near to these have been numerous fragments of petrified Lizards, Tortoises, and sometimes Crabs, in a most perfect state of preservation. Whatever changes however may have taken place in the sea or land to occasion the above phænomena, it is quite natural to suppose that they happened long before the creation of man, since not the smallest remains have ever been found of human bones petrified, in any country, those which have been found in the rocks at Gibraltar (and once supposed to be human) are found enclosed in silicieous earth, but are known by comparison, to belong to some of the Ape or Monkey Tribes.

[This page was blank in the original]

Drawn by G. Perry   Engraved by T.L. Busby

BULIMUS PHASIANUS.

Pub.ᵈ by J. Stratford, Nov. 1810.

PLATE 43

[314]

CONCHOLOGY.

Genus---BULIMUS.
Species---BULIMUS PHASIANUS.

Character---Shell univalve, spiral, the spire and body gibbous; the summit mamillary or rounded, having no beak or rostrum ; the cheek joined to the base of the body by an undulated curve, the form of the left side of the mouth arcuated.

THIS elegant shell, so attractive in its form and colour, is a native of New Holland and Van Diemen's Land, and may be worthily classed with the finest productions of the ocean. In a former number of the ARCANA, we have pre- sented our readers with a representation of the Bulimus Zebra, differing from the present one in having a shorter spire, and a more curvated mouth. The Bulimus Phasianus or Pheasant Shell, so called from its marks resembling those of a Pheasant, is generally about five inches long, of a gently undulated shape and its colour of a rich red tint, verging to a brown, the mouth of a pale blue. The whole of the body, cheek, and spire, are variegated with red streaks and shadows placed above each other, pyramidically, in the most elegant gradation, as if on purpose to catch the eye of the painter, or the admirer of nature. This shell also varies much in the different individuals as to its colour, being sometimes found of a light or dark brown, sometimes olive or very pale, which circumstance perhaps makes them more interesting to the collector, and if well preserved and large, the price they bring is generally very high, from two to three guineas each. The Genus Bulimus has been well elucidated by Brugniere, an eminent French Conchologist, and what is very remarkable, it had been placed with the

Genus Helix which it resembles, in having no beak at the base of the mouth, but in other respects differs very much. There is also a smaller species of the Pheasant Shell than that which we have described, it is generally only one inch and a half in length, the marking of the pattern is much closer, and the spire ends rather more abruptly. These shells are said to be frequently found in the fresh water rivers as well as in the sea, and sometimes are found adhering to the branches and leaves of trees which hang over the stream; if so, this certainly does in some degree join them to the character of the snail, found in our rivers in England, of which we have one or two species of a singularly pointed form, with a lengthened mouth, and which have not yet been described in any recent work of British Conchology.

Of the Bulimus Genus, very few shells have hitherto been found in the European Seas, being generally confined to the more torrid or Southern Regions of the Globe.

[This page was blank in the original]

CERITHIA CŒRULEA.

Drawn and Engrav'd by T.L.Busby

Pub.d by J. Stratford Nov. 1810.

PLATE 44

[318]

Genus--CERITHIA; OR, CREEPER.
Species--CERITHIA CŒRULEA.

Generic Character.--Body and head tapering in form, the bill arched, long, incurvated downwards, gradually acuminated and angular, sharpened at the end; at the base of the bill a nasal opening; three front toes equal, one hinder toe longer than the other three, head much flattened, tail bifurcate.

THE genus Cerithia, or Creeper, forms a curious and interesting family, which naturally stands contiguous in its characters to the Humming Bird, or Trochilus, so remarkable for the radiance and splendor of their colours and for the smallness of their bodies. The French writers seem to have described the genus Cerithia under the name Colibri, although there is perhaps a difference upon which a generic distinction might be formed, if the characters could be more minutely ascertained. At present, therefore we shall consider the Cerithia and Colibri as the same family, for the sake of convenience of arrangement, these birds not being yet sufficiently identified and understood. The Trochilus, or Humming Bird, differs from the Cerithia in the bill, which is more parrallel in its form and sometimes even thickened at the end and more suddenly terminating. There is a peculiarity also in the toes of the Humming Bird, the hinder toe being of the same length as the others. In the rich splendor of their plumage, the Humming Birds are perhaps unrivalled, and in respect to their bright metallic lustre, when examined by certain particular lights, they have been distinguished by the names of the precious

z

stones, as the sapphire, emerald, topaz and ruby, suitably to the characteristic splendor of their varying and inimitable shades.

The Cerithia Cærulea, which we are now about to describe, is in no respect inferior to its relatives the Trochili, or Humming Birds, consisting of three colours, which predominate in the head, wings and body. The top of the head is exactly of the colour of a beautiful turquoise stone, or to speak more plainly, is of a greenish sky-blue. The back and wings variegated with a blue, black and brown colour, the tail brown, and ending in four circular divisions of equal length. Two curious streaks are inserted upon each side of the tail, the scapulars at the bottom of the back, blue striped with yellow; the tufts of the thighs blue, and the legs yellow. The bird altogether has something of the appearance of second mourning from the grey effect which arises from the blue and black shadows; the whole length is about four inches and a half, and it is a native of the Brazils and Mexico. Its elegantly-turned neck, joined to the symmetry and lightness of its form, serve to mark it as one of the greatest favourites of nature, for the purposes of the painter or designer, and the bill is so nicely ballanced to its size as not to appear too long or preposterous for the rest of the body, which is the case with many of the birds in the warmer climates, and its superiority in this and other respects will no doubt recommend it to our numerous readers.

Delineated from the Liverpool Museum.

Extracts from Phillips's Account of the Customs and Arts of the Natives of New Holland.

FROM the very extraordinary shyness of the inhabitants of New Holland and Botany Bay, but little addition has been made to the short, yet faithful account of their manners attained by the observations of Captain Cooke, and the earliest circumnavigators of this curious and interesting country. The whole indeed that can be known of a people amongst whom civilization and the arts of life have made so small a progress, must amount to very little in the enumeration. They seem not to have discovered the manufacture or uses of large nets, as the New Zealanders have, but some small nets have been obtained from them, the construction of which is very curious. The twine of which they are made, appears to be composed of the fibres of a plant resembling flax, with very little previous preparation, it is very strong, heavy, and so admirably well twisted as to have the appearance of texture of the best whip cord. Some of the sailors had obtained lines of their manufacture, which were made from the fur of some animal, and others that appeared to be of cotton ; the meshes of their nets, very artificially inserted into each other but without any knots. At a small distance they have exactly the appearance of our common nets, but when they are closely examined, the peculiar mode in which the loops are arranged, is found to be different and very remarkable. Some ladies who have inspected one of these nets lately imported, declare that it is exactly on the same principle as the ground of point lace, except that it has only one turn of the thread instead of two, in every loop. These nets appear to have been used either as a landing net, or for the purpose of carrying the fish when taken. They have also small hoop nets in which they catch Lobsters and other sea cray fish. Their canoes are small and narrow, and sometimes joined together with cross boards; they also use a short paddle or oar, rounded at the lower end and flattened. These savages are tall and thin, their mouths large, of a visage dark and disgusting; they have few ornaments for their persons, except such as are impressed upon the skin itself or laid on in the manner of paint. The men keep their beards short, as it is thought by scorching off the hair, and several of them at the first arrival of our people, appear to take great delight in being shaved. They sometimes hang in their hair the teeth of dogs and other animals, the claws of lobsters, and several small bones, which they fasten there by means of gum; but such ornaments have never been seen upon the women, though they seem not to make any rational attempt at cloathing themselves; they are by no means insensible of the cold, and appear very much to dislike rain. The Governor therefore being convinced by these circumstances, that cloathing would be very acceptable to them, if they could be induced to come sufficiently amongst the English to learn the use of it, ordered a supply of frocks and jackets to be made long and loose to serve either for men or women. The bodies of these people smell very strangely of oil, and the natural darkness of their colour is much increased by dirt. But although in these points they shew so little delicacy, they are not without emotion of disgust when they meet with strong effluvia, to which their organs are unaccustomed. One of them having touched a piece of pork, held out his finger for his companion to smell, with strong marks of distaste. Bread and meat they seldom refuse to take, but generally throw it away soon after. Fish they always accept very eagerly. Whether they use any particular rites of Burial is not yet known, but it seems from the following account evident that they have the custom and practice of burning their dead.

The ground had been observed to be raised in several places, like the ruder kind of graves of the common people in England, Governor Phillips ordered some of these barrows to be opened. In one of them a jaw-bone was found not quite consumed, but in general they contained only ashes. From the manner in which these ashes were disposed, it appeared that the body must have been laid at length, raised from the ground a few inches only, or just enough to admit a fire under it, and having been consumed in this posture it must then have been lightly covered over with mold; fern is generally spread upon the surface with a few stones to keep it from being dispersed with the wind. These graves have not been found in very great numbers, nor ever near their huts.

The natives of New South Wales, though in so rude and uncivilized a state as not even to have made an attempt towards clothing themselves, notwithstanding that at times they suffer from the cold and wet, are not without notions of sculpture. In all these excursions of Governor Phillips, and in the neighbourhood of Botany Bay and Port Jackson, the figures of animals, of shields and weapons, and even of men, have been seen carved upon the rocks, roughly indeed, but sufficiently well to ascertain very fully what was the object intended. Fish were often represented, and in one place the form of a large Lizard was sketched out with very tolerable accuracy. On the top of one of these hills the figure of a man in the attitude usually assumed by them when they begin to dance was executed in a still superior style. That the arts of imitation and amusement should thus in any degree precede those of necessity, seems an exception to the rules laid down by theory for the progress of invention. But perhaps it may better be considered as a proof that the climate is never so severe as to make the provisions of covering or shelter a matter of absolute necessity. Had these men been exposed to a colder atmosphere, they would doubtless have had cloaths and houses, before they attempted to become sculptors. The country explored in some of the inner parts was so good and fit for cultivation, that the Governor resolved to send a detachment to settle there as soon as convenient. The natives however who know not how to avail themselves of the fertility, are still very numerous in the inland country, and it is wonderful how they subsist. Near to one of their huts, the bones of a Kangaroo were found and several trees were seen half burnt, and it seemed evident that the natives had fled at the approach of the English party, but so effectually did they conceal themselves that not one was seen. The huts seen here consist of single pieces of bark about eleven feet in length and from four to six in breadth, bent in the middle, while fresh from the tree and set up so as to form an acute angle, not a little resembling cards set up by children.

In the few we visited, some spears were found, and it was conjectured that the use of these structures might partly be to conceal themselves from the animals for which they must frequently lie in wait. They may also afford shelter from a shower of rain to one or two who sit or lie under them. The men are distinguished by different marks, some of them want the tooth of the right front jaw; Governor Phillips having remarked this, pointed out to them that he himself had lost one of his front teeth, which occasioned a general clamour. There is sometimes a perforation in the cartilage which divides the nostrils and the strange disfiguring ornament of a long bone or stick thrust through it, was now observed as described by Captain Cook, and the same appellation of spritsail yard was applied ludicrously to it by the sailors.

But several very old men were seen who had not lost the tooth, nor had their noses been prepared to receive that grotesque appendage, probably therefore these are marks of distinction; ambition must have its badges, and where cloaths are not worn the body must be compelled to bear them. The women seemed in some instances to have one of their fingers mutilated, by cutting off the first and second joint. They made no attempt to secret themselves, nor seemed impressed with any idea that one part of the body more requires concealment than another, yet there was a shyness and timidity among them which frequently kept them at a distance. They never would approach so readily as the men, and sometimes would not even land from their canoes, but made signs that what was offered them should be given to the men. We are not yet sufficiently acquainted with the manners of the people, to decide whether this reserve proceeds from the fears of the women or from the jealousy of the husbands, by whom they are evidently kept in great subordination.

One of their modes of fishing was now observed; their hooks were made of the inside of a shell resembling mother of pearl. When a fish which has taken the bait is supposed to be too strong to be drawn up with the line, the canoe is paddled to shore and while one man gently draws the fish along, another stands prepared to strike it with the spear, in this attempt they seldom fail. When the southern branch of Broken Bay was first visited, the getting round the headland that separates the branches, was very difficult on account of the squalls of wind accompanied with rain. An attempt was made to land where there was not sufficient water for the boat. During this transaction an old man and a youth were standing on a rock where the boat was trying to approach. Having seen how much our men had laboured to get under land, they were very solicitous to point out the deepest water. Afterwards they brought fire, and seemed willing to render any service in their power, two of the officers suffered themselves to be conducted by the old man to a cave at some distance, but declined going in, though he invited them by all the signs he could invent. This was rather unfortunate, as the rain was falling very violently and the cave was found next day large enough to have sheltered the whole party. The old man afterwards assisted in clearing away the shrubs and making preparations for the party to sleep on shore, and next morning he was rewarded with presents for his friendly behaviour, and a hatchet seemed to be one of the most acceptable gifts he could receive The rain which was almost constant, prevented the Governor from returning by land and of course from making further efforts of conciliation and regard, who seemed however to have no sense of the immorality of thieving until put in mind of it by the notices of some of the crew, upon which the things which had been pilfered were readily given up, without any observations by their or our party.

[This page was blank in the original]

Drawn by G.Parry

Engrav'd by T.L.Busby.

HIPPOCAMPUS.

Pub.^d by J.Stratford,Decr 1.1810.

PLATE 45

Genus---SYNGNATHUS; OR, HIPPOCAMPUS.
Species---ERECTUS.

Character—Animal having a head formed like a Horse, body jointed like armour, the fins placed on a pedicle, irregular in their number and position, no caudal or terminating-fin to the tail.

IN a former number of the ARCANA (for May) we imparted to our readers a new species of this curious genus of sea animals, and which was of a larger size, and much longer form than the present Hippocampus, although not quite so round or broad in the body. The form of the fins also were found to be materially different, and much more numerous. The rarity however of the former animal is greater than in the present instance, yet the whimsical fancy, if we may so call it, is equally displayed in the pattern of each, and it is not, perhaps, impossible that painters and sculptors may first have borrowed the idea of a mermaid, from the specimen now before us, and excepting the head and the want of arms and hands, it strongly reminds us of that object.

The Hippocampus Erectus is a native of the American Seas, and of the coasts adjacent to Mexico and the West Indies; its size varies from seven inches to nine, in various specimens, and which, perhaps, is distinctive of the different sexes, the male being the smallest. The head has very much the resemblance to that of a Horse, in the way represented by the ancient sculptors of Greece and Rome, the same similitude is kept up wonderfully in the proportion and form of the neck; the organs for hearing being placed

A a

at the back part of the neck; and forms externally, an angular opening. The colour of the body is of a pale amber, shaded with brown, and which is divided into ribs transversely placed, and continued in a closer manner upon the neck and tail; the mouth is truncated and without teeth, and has two small horns standing upon the forehead immediately above the eyes. Higher up and projecting from the crown of the head are two pointed tubercles, and one below, fixed upon the under jaw; the back is invested with a spreading fin, which is filamentous and pointed downwards. In the front part of the abdomen, are placed two small circular fins curvated, and these are all which the animal seems to possess.

In the infinite varieties which occur in the different kingdoms of animals, fishes, birds and insects, we have had frequently our attention drawn to those intergenera, or connecting links, which unite by analogy, two different tribes of beings. Thus the Bat exhibits a gradation, being placed between the bird and the quadruped; the flying fish, endowed with a power of moving through the air, joins the characters of the fish and bird, and the present specimen seems to unite the qualities of fish and insect, its covering being divided into partitional segments, yet without any fin to its tail; it still has a sufficient analogy, in its situation and habits to be reckoned by a superficial observer, a fish, but a difference in its form from all fish, is observable thoughout the whole of this most curious animal.

[This page was blank in the original]

Drawn by G. Perry.

ARCUATUS CŒRULEUS.

Engraved by T.L. Busby.

Pub.d by J. Stratford Dec.r 1.1810

PLATE 46

Genus---PAPILIO.   *Division*---ARCUATUS.

IT was formerly the method adopted by some eminent naturalists, to class the Butterfly-tribes by placing
them under five principal divisions, or families, and these
again into sub-divisions, according to the spots, the transparency of the wings, or the colour or form of the upper
wings. This was the custom of LINNÆUS and FABRICIUS,
and is found now to be entirely wanting in decision and
preciseness of terms, for in many instances, the spots vary
in number and colour, even in the same individual, from
difference of sex, or of food and situation.

The terms of Rurales, Danai, Equites, Nymphæ, &c.
are so confused and contradictory to each other, so little
explanatory of their nature and qualities, and drawn from
such opposite and unmeaning circumstances, added to the
pompous and ridiculous names of the Greeks and Trojans,
that every judicious Papilionist naturally wishes for a new
arrangement in that difficult, yet interesting tribe of insects.
To enter into the full disquisition of the errors and obvious
contradictions of such a mode of classification, would be
inconsistent with the conciseness of our present plan in this
work, as we wish rather to draw the attention of our readers
to the general observations of nature, than to abstruse distinctions of difficult terms. We would therefore suggest,
in preference to the above, a method adopted by an ingenious naturalist of the present time, who proposes to include
all the Papilio-tribes which are found in nature, under
certain definitional characters, taken entirely from the forms
of the wings, which will impart to us in every word, some
particular character of its shape and proportion.   This,

undoubtedly, will be found more useful and clearly understood than to take the definitions from the spots, colour, or
transparency of the wings, for these are vague and uncertain
and perpetually varying, whereas the external outline
(which is a grand distinction) is always found to be constant
and persisting.

According to this new division, the present Butterfly
comes under the character called Arcuatus, or bow-shaped,
having in the lower part of the outline an undulated shape,
by the junction of the upper and lower wings, which makes
it resemble a bow. The Orbati have their wings very much
extended, and all of them rounded, as described in the last
number. The Caudati have a long tail, projecting from
each lower wing, and these form a very large family of
foreign Papilios, although but very few of them are found
in England. The Excelsi have their upper wings spread
out, rounded, and very much lifted up. The Cuspidati
have the outermost corners of the upper wings cut off in
an angular form. The Muscarii have wings resembling the
common fly, and which are also transparent.

It appears likely, from the convenience and perspicuity
of the above arrangement, that it may very probably super-
sede the necessity of all the former ones, and illustrate
clearly, by the most exact definitions, this most beautiful
and interesting branch of the animated creation.

[This page was blank in the original]

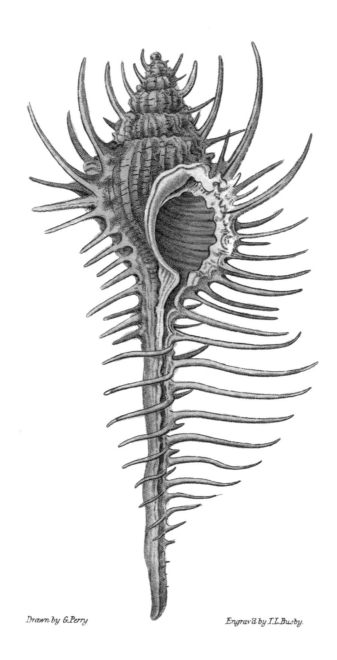

Drawn by G.Perry

Engrav'd by T.L.Busby.

ARANEA GRACILIS.

Pub.d by J.Stratford Dec.1.1810.

PLATE 47

CONCHOLOGY.

Genus--ARANEA. Species--ARANEA GRACILIS.

Character.—Shell univalve, spiral, the spire and body short and rounded, the beak long and armed with a triple row of spines, the mouth undulated and labiated. The body, spire, and beak invested with a triple accumulation of curved and pointed integuments, open at the base.

THE curious and graceful shell of which we are now about to present the resemblance to our readers, was classed by that great naturalist LINNÆUS, along with the Murices, by the name of Murex Tribulus Minor; upon a further investigation, however, of its form, it seems more properly to form a genus of itself, of which, about twenty different species are at present known, some of which are three times as large as the present. By some of our later writers upon Conchology, it has been called by the name of Venus's Comb, or the small Thorney Woodcock, from its supposed resemblance to a Woodcock's head and bill.

Of those shells which are denominated the branched species, the Triplex and the Aranea are the most remarkable, the distinction which exists between them, has been remarked in a former number, where several of the Triplices have been already described. The length of the shell of the Aranea Gracilis is generally from five to six inches, and exhibits a striking and pleasing object, as to the elegance, lightness, and intricacy of its parts. The number of its curved spines or thorns, amounts in the whole to ninety-five, all of different lengths, and placed each of them at various distances, in the most curious and agreeable variety. The beak is elongated and tapering, and at the bottom slightly bent on one side, the spire short and ending in a round tubercle at the top. The colour of the whole shell is generally of a pale amber tint, inclining to a red, the mouth sometimes white, red, or brown, richly streaked with circular lines. The most elegant specimen of this shell which we have hitherto seen, is that which was in the late Mr. CRACHERODE's collection, and now deposited in the British Museum, the comparative value being appreciated by the number, length and preservation of the spines. The shells which we have hitherto delineated, have, many of them, been remarkable for a boldness of outline and richness of colouring, from which however, the Aranea differs most materially, recommending itself chiefly by a graceful lightness of form, with a great intricacy and diversity, which our sublime author the late Mr. BURKE, as well as our illustrious HOGARTH, have described, as being principally necessary to the impressions of beauty. Such is the astonishing variety of character in each part of individual nature, from which, undoubtedly, all the principles of artificial taste and beauty were traced and designated by the ingenious and active powers of man.

[This page was blank in the original]

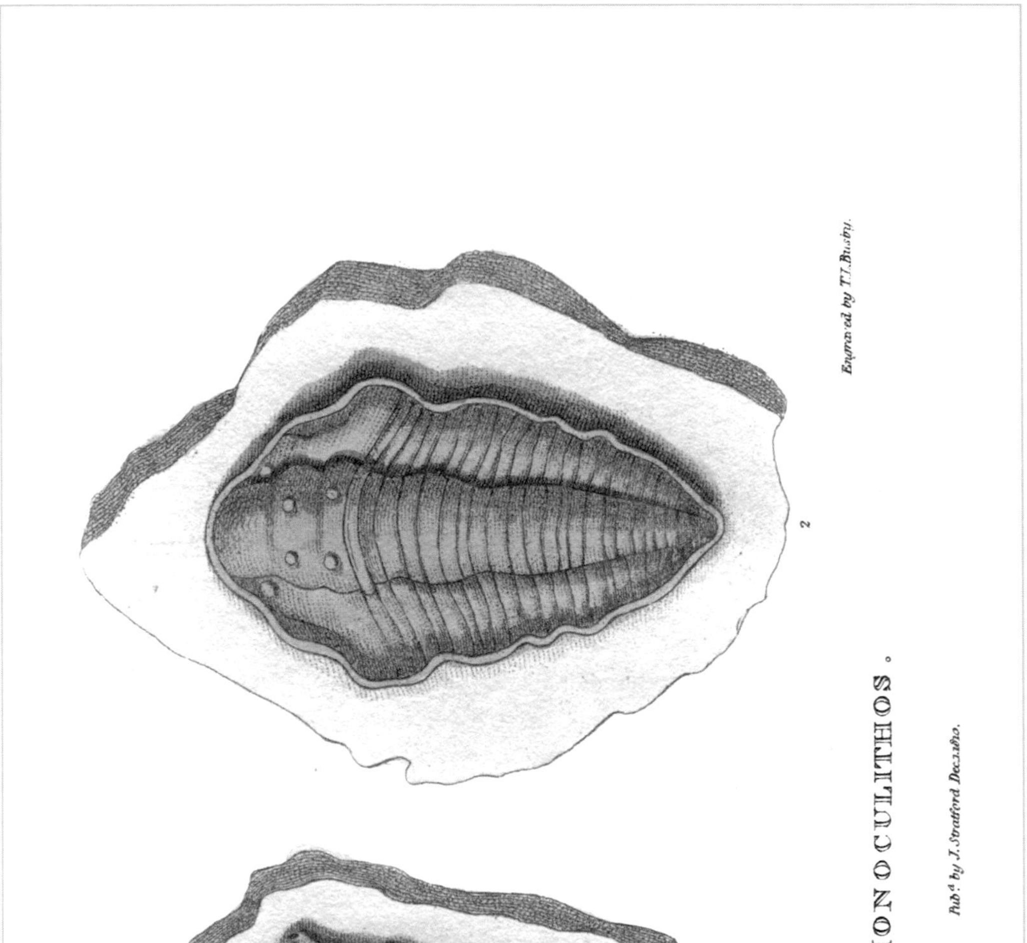

MONOCULITHOS.

Drawn by G. Perry.

Pub.d by J. Stratford Dec.r 1810.

Engrav.ed by T.L.Busby.

PLATE 48

[340]

FOSSILIA. *Family*—NANTES, or SWIMMERS.
Genus—MONOCULITHOS.

Species Two—1. POLYMORPUS. 2. HEXAMORPHUS.

THE curious circumstances of the Dudley fossils, have already engaged a considerable portion of our attention in the preceding number, and at the same time we endeavoured, as far as our brief limits would allow, a short but general account of some pecularities in the fossil kingdom, and we shall now resume, being a subject hitherto not sufficiently discussed, so far as relates to the animal-remains of our own kingdom. It perhaps has not escaped the observation of the reader, that a great number of the polished marbles used in the decorations of art, contain a numberless infinitude of fossil shells, and of animal and vegetable exuviæ. The marble quarries of Kilkenny, in Ireland, contain an amazing number of petrified shells inclosed in a black ground, and which, when cut across, resemble the circular slices of onions, that these are not of the Cardium, or Cockle-kind, is obvious, and specimens of the Cornu Ammonis, Argonauta, and of many different patterns of the Monoculithos, as well as the Tubeporite and Madrepore, are frequently inserted, all attesting their marine origin. In Derbyshire, Scotland, and Wales, even upon the most inland vallies and mountains, the same objects are exhibited in strata of limestone, chalk, or clay, attesting strongly the universality of the flood.

If any circumstance were wanting to prove to our senses their marine origin, nothing can be stronger than one of the present instances of Monoculithos Polymorphus, No. 1, where a small fossil-shell of the Mya genus is found attached to the side of the stone, containing the fossil. This specimen is remarkable also for the numerous tubercles which invest the upper part, or head, and two appendages, or arms, which hang down on the outside, also tuberculated; similar traces, or arms, are also visible in the large one, published in a former number. The tail of No. 1 is curiously fimbriated and adorned with tubercles and regularly diminishing.

No. 2, is the Monoculithos Hexamorphus, and has only six tubercles upon the head, the rest of the body, and what is generally called the tail, is much less ornamental in all its parts than the former. These specimens are generally of a flat shape and character, and the under side seldom distinguishable from the upper, and therefore not to be seen completely all round. There is, however, a striking likeness in the general analogy of the parts, which may perhaps lead to a further investigation of their anatomical structure. It may also perhaps be necessary in this part of our subject to add the conjecture of an ingenious modern naturalist, who supposes them to be the larvæ, or chrysalis of some large Moth, or Insect, in a withered or contracted state.

Extracts from the Travels of Mr. Mungo Parke in the Interior of Africa, containing an Account of the River Niger.

ARRIVING at Wassiboo, one of the principal towns in the neighbourhood of the kingdom of Bambarra, I met with eight fugitive Kaartans, who offered to accompany me to Satile, and I acquiesced in their proposals; at day-break we set out, and travelled with uncommon expedition until sun-set; we stopped only twice in the course of the day, once at a watering place in the woods, and another time at the ruins of a town formerly called Illa-Campe (the Corn Town.) When we afterwards arrived in the neighbourhood of Satile, the people who were employed in the corn fields seeing so many horsemen took us for a party of Moors, and ran screaming away. The whole town was alarmed, and the slaves were seen in every direction driving the cattle and horses towards the town. It was in vain that one of our company galloped up to undeceive them, it only frightened them the more, and when we arrived at the town we found the gates shut and the people all under arms. After a long parley we were permitted to enter, and as there was every appearance of a heavy tornado, the black governor allowed us to sleep in his grounds, and gave us each a bullock's hide for our bed. Early in the morning we again set forward, the roads were wet and slippery, but the country was every where beautiful, abounding with rivulets, which were increased by the rain into rapid streams, we shortly afterwards arrived at Moorja, a large town famous for its trade in salt, which the Moors bring here in great quantities, to exchange for corn and

B b

cotton cloths. As most of the black people here are Maho-
medans, it is not allowed to the Caffirs to drink beer,
which they call neo-dollo (corn spirit) except in certain
houses. In one of these I saw about twenty people sitting
round large vessels of this beer with the greatest conviviality,
many of them in a state of intoxication. As corn is plentiful, the inhabitants are very hospitable and liberal to
strangers, and I believe we had as much corn and milk
sent us as would have been sufficient for three times our
number, and though we remained there two days, we
experienced no dimunition of kindness and regard.

We reached the next village, called Datliboo, and
passed a caravan of travellers with corn paddles, mats,
and other household utensils, returning from the town of
Sego. We continued our journey, but having had a light
supper the preceding night, we felt ourselves rather hungry
and endeavoured to procure some corn at this village, but
without success. The towns were now more numerous,
and the land that is not employed in cultivation, affords
excellent pasturage for large herds of cattle, but owing to
the great concourse of people daily going to and returning
from Sego, the inhabitants are less hospitable to strangers.
My horse, in this part of my journey, being very much
fatigued, I was walking barefoot and driving my horse,
when I was met by a caravan of slaves, about seventy in
number, coming from Sego. They were tied together by
the necks with thongs of a bullock's hide twisted like a
rope, seven slaves upon a thong, and a man with a musket
between every seven. Many of the slaves were ill conditi-
oned and a great number of them women; they were to
proceed by the Ludamar and the great desart to Morocco.
At eight o'clock we departed from Doolinkaboo, and stopped
at a large village; hearing that two negroes were going
from thence to Sego, I was happy to have their company,

and we set out immediately. About four o'clock we stopped at a small village, where one of the negroes met with an acquaintance, who invited us to a sort of public entertainment, which was conducted with uncommon propriety. A dish made of sour milk and meal called sinkatoo, and beer made from their corn was distributed with great liberality, and the women were admitted into the society. There was no compulsion, each was allowed to drink as he pleased, they nodded to each other when about to drink, and on setting down the calabash commonly said berka (thank you.) Both men and women appeared to be somewhat intoxicated, but they were far from being quarrelsome. We now began to approach the city of Sego, the residence of the King and the capital of Bambarra, to whom the Kaartans, my companions, promised to introduce me.

Aterwards we rode through a marshy valley, and as I was looking out anxiously for the river, one of them called out " geo-ffilli," (see the water) and looking forwards, I saw with infinite pleasure the great object of my mission, the long sought-for majestic Niger glittering to the morning sun, as broad as the Thames at Westminster, and flowing slowly to the eastward. I hastened to the brink, and having drank of the water, lifted up my fervent thanks in prayer to the Great Ruler of all things, for having thus far crowned my endeavours with success. The circumstance of the Niger flowing towards the east and its collateral points, did not however excite my surprize, though I had left Europe in great hesitation on this subject, and rather believed that it ran in a contrary direction. I had made such frequent enquiries during my progress concerning it from negroes of different nations, such clear and decisive assurances, that its general course was towards the rising sun, as scarce left any doubt on my mind, especially as Major HOUGHTON had received similar information.

Sego, the capital of Bambarra, consists properly of four distinct towns, two on the north bank of the Niger, and named Sego-korro and Sego-boo, two on the south called Sego-soo-korro and Sego-see-korro. They are all surrounded with high mud walls, the houses are built of clay, of a square form, with flat roofs, some of them have two stories, and many of them are whitewashed. Besides these buildings, Moorish mosques are seen in every quarter, and the streets, though narrow, are broad enough for every useful purpose where wheel carriages are entirely unknown. From the best enquiries I could make, I have reason to believe that Sego contains altogether about thirty thousand inhabitants. The King of Bambarra resides at Sego-see-korro, he employs a great many slaves in conveying people over the river, and the money they receive (though the fare is only ten cowries for each person) furnishes a very large revenue in the course of the year. The canoes are of a singular construction, each of them being formed of the trunks of two large trees, rendered concave and joined together, not side by side but end to end, the junction being exactly across the middle of the canoe, they are therefore very long and disproportionably narrow, and have neither decks nor masts, they are however very roomy, for I observed in one of them four horses and several people crossing over the river. When we arrived at this ferry, we found a great number waiting for a passage, they looked at me with silent wonder, and I distinguished with concern many Moors among them.

There were three different places of embarkation, and the ferry-men were very diligent and expeditious; but from the crowd of people that had assembled, I could not immediately obtain a passage, and sat down upon the bank of the river, to await a more favourable opportunity. The view of this extensive city, the numerous canoes upon the river, the crowded population, and the cultivated state of the surrounding country, formed altogether a prospect of civilization and magnificence, which I little expected to find in the bosom of Africa. I waited more than two hours without having an opportunity of crossing the river, during which time, the people who had crossed carried information to Mansong the King, that a white man was waiting for a passage and was coming to see him. He immediately sent over one of his chief men, who informed me, that the King could not possibly see me, until he knew what had brought me in his country, and that I must not presume to cross the river without the King's permission. He therefore advised me to lodge at a distant village, to which he pointed, for the night, and said that in the morning he would give me further directions how to conduct myself. This was very discouraging, however as there was no remedy, I set off for the village, where I found to my great mortification, that no person would admit me into his house. I was regarded with astonishment and fear, and was obliged to sit all day without victuals, in the shade of a tree; the night threatened to be very uncomfortable, for the wind arose and there was great appearance of a heavy rain, and the wild beasts are so very numerous in the neighbourhood, that I should have been under the necessity of climing up the tree and resting amongst the branches; about sun-set however as I was preparing to pass the night in this manner, and had turned my horse loose that he might graze at liberty, a woman returning from the labours of the field, stopped to observe me, and perceiving that I was weary and dejected, inquired into my situation, which I briefly explained to her, whereupon, with looks of great compassion, she took up my saddle and bridle and told me to follow her; having conducted me to her hut, she lighted up a lamp, spread a mat for me on the floor, and told me I might remain there for the night;

finding I was very hungry, she said she would procure me something to eat; she accordingly went out and returned in a short time with a very fine fish, which having caused to be half broiled upon some embers, she gave me for my supper. In the afternoon another messenger arrived with a bag in his hands, he told me it was the King's pleasure, that I should depart forthwith from the neighbourhood of Sego, but that wishing to relieve a white man in distress, had sent me five thousand cowries, to enable me to purchase provisions in the course of my journey. Being thus compelled to leave Sego, I was conducted the same evening to a small village about seven miles to the eastward, with some of which my guide was acquainted, by whom we were well received. He was very friendly and communicative, and informed me that the cities of Tombuctoo and Jenne were under the dominion of the Moors. About eight o'clock we passed a large town called Kabba, situated in the midst of a beautiful and highly cultivated country, bearing a considerable resemblance to England. The people were employed in collecting the fruit of the shea trees, from which they prepare a vegetable butter, and which grow in all parts of Bambarra in great abundance; they are not planted by the natives, but grow wild and in clearing wood land for cultivation, every tree is cut down except the shea. The tree itself much resembles an American oak, and the fruit from the kernel of which being dried in the sun, the butter is prepared by boiling the kernel in water, has something of the appearance of a Spanish olive; the kernel is enveloped in a pulp and under a green rind, the butter produced from it, besides the advantage of its keeping a whole year without salt, is whiter, firmer, and to my palate, of a richer flavour than the best butter I ever tasted made from cows' milk; the growth and preparation of this commodity seem to be among the first articles of African industry in this and the neighbouring states, and it constitutes a main article of their inland commerce. We passed, in the course of the day, a great many villages, chiefly inhabited by fishermen, and in the evening arrived at Sansanding, a large town containing as I was told, from eight to ten thousand inhabitants; this place is much resorted to by the Moors, who bring salt from Beroo and other Moorish countries, where on account of the want of rain, no cotton is cultivated; we rode near the town and river, passing by a creek or harbour, where I observed twenty large canoes, most of them fully loaded and covered with mats to keep out the rain, and in short several others arrived with passengers. The ensuing night I slept at Sibili, from whence, the next day I arrived at Nyara a large town at some distance from the river, and the governor very civilly sent his son to shew me the road to Modiboo, which he assured me was at no great distance; we rode in a direct line thro' the woods, but in general went forward with great circumspection, my guide frequently stopping and looking under the bushes; on my enquiring the reason of this caution, he told me that lions were very numerous in that part of the country, and frequently attacked persons travelling in the woods; while he was speaking my horse started, and looking, I observed a large animal of the Cameleopard kind standing at a little distance, the neck and fore legs were very long, the head was furnished with two short black horns turning backwards, the tail, which reached down to the ham joint, had a tuft of hair at the end; the animal was of a mouse colour, and it trotted away from us in a very sluggish manner, moving its head from side to side to see if we were pursuing it.

Shortly after this, as we were crossing a large open plain, where there were a few scattered bushes, my guide, who was a little before me, wheeled his horse round in a moment, calling out something in the Foulah language, which I did not understand, I enquired in Mandingo what he meant, " wara, billi, billi," (a very large lion) said he, and made signs for me to ride away; my horse was too fatigued, so we rode slowly past the bush from which the animal had given us the alarm, not seeing any thing myself however, I thought my guide had been mistaken, when the Foulah suddenly putting his hand to his mouth, exclaimed " soubah an Allahi," (God preserve us) and to my great surprise, I then perceived a large lion at a short distance from the bush with his head couched between his fore paws, I expected he would instantly spring upon me, and instinctively pulled my feet from the stirrups, that my horse might become the victim rather than myself; but it is probable the lion was not hungry, since he quietly suffered us to pass, though we were fairly within his reach; my eyes were so rivetted upon this sovereign of the beasts, that I found it impossible to remove them until we were at a considerable distance.

[This page was blank in the original]

Drawn by G. Perry.   Engrav'd by T.L. Busby.

CAMELEOPARD.

Pub.d by J. Stratford, Jan.1.1811.

PLATE 49

Genus-CAMELUS.
Species--CAMELOPARDALIS; or, TORTOISE-COLOURED ANTELOPE.

Character.—Horns short and small, united by the skull; the neck and legs very long; the body short and spotted closely with various markings, octagonal, oval, or square; the ears placed upon the neck; tail short, ending in a hairy extremity; the feet cloven and obtusely pointed; the chest high and projecting; found only in the Torrid Zone.

THE Camelopard is an animal which is for the singularity of its form, and immense size, justly attractive of the admiration of mankind, and presents to all those who are deeply interested in Natural History, a strong and wonderful instance of the great variety, which the author of nature has spread through all his works. The head has a striking resemblance to that of a Horse, but differs from it essentially in having a high boney process, shaped like the keel of a ship, and placed in the centre of the forehead between the eyes; the horns are small and rounded at the points and covered with short hairs at the ends. The legs are beautifully taper, as well as the neck, and whether we consider the variety of spots; their mathematical shapes, or the stately contour visible in its general character, we may certainly consider it as one of the most majestic animals of the creation. The Horse indeed presents to the eye of the spectator a different set of proportions, to which our judgment has become more familiarized, and is of course more connected with the ideas of utility and intellect than the former, or any other animal at present discovered.

c c

ZOOLOGY.

The singularity of the Kanguroo and its numerous species, arises from the extreme shortness of the fore legs and length of the hinder ones. In the present animal, this system is reversed, the fore legs appearing by much the longest. The Camelopard is a perfectly harmless animal, and subsists in the middle and southern parts of Africa, by grazing, and also by feeding on the young branches of the trees, for which his long neck is admirably calculated. He is supposed to be quite incapable of domiciliation and his great size would render him probably a most inconvenient animal even if tamed. The present delineation was taken from a fine and noble specimen preserved in Mr. BULLOCK's Museum, and which was shot by an English Gentleman near the Cape of Good Hope. It is found to be the largest specimen ever brought to England, being seventeen feet in height; but no description whatever can impart the idea of vastness, which the sight of the animal itself must always inspire in the spectator. He traverses his native plains, when alive, in herds of twenty or thirty together, and when grazing is said to bring his head very low between his fore legs in a striding posture. When pursued by dogs or men, of both of which he is much afraid, he commences a brisk ambling trot, and which is afterwards increased in velocity, leaving his pursuers far behind. The Camelopard however, though now very scarce, was well known to the ancient Romans, having been frequently exhibited by them in their Circus and publick games, for the gratification of the populace, who delighted much in these kind of exhibitions.

Many various observations have been made by different naturalists, respecting the comparative anatomy of the Camelopard, and it has been observed by VAILLANT, that the same protuberance occurs in this animal, which is found upon the back of the Camel and Dromedary. Even in the Lama and Vicuna, which are found in South America, described in a former number of the ARCANA, the same elevated rising is observed. In all these animals the hump or sudden rising of the back, differ considerably; in the Camel it is very large, and placed in the centre; in the Dromedary divided into two prominences, as if for the purpose of placing the load with safety, and in the Lama and Vicuna it is much less obvious. In the Camelopard the same circumstance exists, only placed closely to the upper part of the shoulders. This protuberance is to be considered as a fleshy florescence or tumour, and not connected with any enlargement of the bone in that part.

The second singular point in the formation of the Camelopard, is the circumstance of the horns which grow upon the top of the head, proceeding from a raised bony process, which is elevated and higher than the rest of the skull. It appears also very plain that the beast has not the power of casting them, like the Stag, the Elk and other ruminating animals, but that they are to be considered as a part of the skull itself, and dating their existence from the very birth of the creature.

The account of PATTERSON, who in his Botanical Tour in the neighbourhood of the Cape of Good Hope, shot one of these curious animals, agrees with the general account as to their mildness and timidity. The individual which he killed and which is now in the Museum of the late DOCTOR HUNTER, in Lincoln's Inn Fields, was about fifteen feet only in height, and may therefore be properly supposed to have been a young one, or a female.

ZOOLOGY.

The next singularity which we shall notice in this astonishing quadruped is the great number of spots which amount in the whole to about four hundred. These are of different sizes and mathematical forms such as the square, octagon, circle, oval, pentagon and hexagon almost as if they had been drawn by a mathematician with his compasses and square.

The public will no doubt consider themselves as much indebted to the liberal generosity of Mr. BUL-LOCK, in procuring the view of this gigantic animal which is certainly the tallest in the known world. At a considerable expence and exertion, he has obtained it from abroad and finally preserved it here: it will long remain, an example of a variety in the quadruped divisions, though once supposed by sceptical minds to have been wholly invented by fabulous travellers.

[This page was blank in the original]

C.Hirst dd.

T.L.Busby sculp.

Red Phalarope

Pub.d by J. Stratford. Jan.1.1811

PLATE 50

Order--GRALLÆ; OR, WADERS.
Genus---TRINGA; OR, PHALAROPA.
Species--TRINGA RUBRA. *Variety.*

Generic Character.—Bill slender and incurvated, either
    upwards or downwards, legs rather lengthened,
    the toes divided and partly webbed or scalloped,
    the eyes placed very much backwards and up-
    wards in the head, the mandibles pointed and of
    equal length.

THE Grallæ genus are one of the six natural divisions of birds, laid down by LINNÆUS, and meant by him
to include the Herons, Curlews, Plovers, and other water
birds, having in general three long toes in front, and one
short one recurved behind, some exceptions however occur
in certain instances, as their feet being webbed or scalloped,
as in the bird now before us; for in some instances, the
water-birds of this division seem to be half webbed. It
seems therefore that in this part of natural history, a nicer
distinction should be drawn between these families of birds,
called Grallæ, and which might easily be done from the
particular form of the feet.

We have received from Mr. PRIESTNALL, of Stockport, a drawing of the Red Phalarope, which is conceived
to be a very rare and singular variety of that curious bird,
and which is described by him in the following manner.
" The Red Phalarope was shot near Stockport, in the
winter of 1806 ; it weighed two ounces only : the bill was
black, slender, and bent a little towards the tip, a dusky
stripe passed below the eye to the back of the head, where

it is joined to a reddish stripe which falls down to the neck, the chin and throat white, with small brown spots, the hinder part of the head, neck, and back of a ferruginous colour, and the wings reaching to the end of the tail. The breast is very white and faintly spotted with brown, the legs of a beautiful blue colour, the toes scalloped and serrated, the nails black." In the entertaining and useful work of Mr. BEWICK, upon the Land and Water Birds in England, we suppose that he alludes to this bird in the following words. " Tringa Hyperborea, or Red Phalarope : the bill is slender and straight, about an inch long and bent a little downwards at the tip, this species is very rare in England."

[*Note of the Editor.*] Mr. BEWICK's remarks respecting the scarcity of this bird, may be applied to a great many others which are found in England, we shall therefore feel ourselves highly indebted to those correspondents, who are possessed of a lucky opportunity of adding to this desirable knowledge of the varieties of nature, as there is still an ample field for discoveries of the different tribes of insects, birds, fish, fossil remains of animals, &c. which will alone impart a pleasing novelty to a work professing to exhibit the secrets of nature ; and it is almost unnecessary to say that all such communications will be most thankfully received. If accompanied by accurate coloured drawings, faithfully representing the various objects of description, our pleasure and advantage will be considerably encreased, and a rich harvest of knowledge and investigation will amply repay our time, labour, and exertions.

Engrav'd by T.L.Busby

Drawn by G.Perry.

PHALÆNA

Pub.d by J.Stratford, Jan.1.1811.

PLATE 51

Distinctive Genus---PHALŒNA.
Division---ARCUATA VITREA; OR, ATLAS MOTH.

THE fanciful powers of the mind of the great Lin-næus, have no where been more plainly exemplified than in his various descriptions of the insect tribes. The name of Atlas has been therefore given by him to the moth which is here represented, and who was reported by the ancient mythological writers, to be so strong as to be able to carry the whole world upon his shoulders, in which attitude he is generally represented by the ancient sculptors. Atlas is said also to have been a great astronomer and to have examined the courses of the planets and stars, from that celebrated mountain in Africa which still bears his name. The present moth being of a larger size and strength than most others of his tribe, might perhaps be his reason for that distinctive name.

The Phalæna of our present number, is a native of South America, and like most of that tribe of insects, is of a predominent brown colour, in opposition the Papilios, which consist in general of wings more variegated in their tints. Its form is arcuated, or shaped like a bow, and presents a most graceful and elegant outline on every side. On the middle of each upper and lower wing, is placed a very remarkable spot angular, of a whitish brown colour, and of a transparent appearance similar to glass or tale. We prefer therefore the name of vitrea, or glassy, for the specific description, as intimating more plainly those peculiar marks.

The wings of all the Papilios and Phalænas already discovered are composed of a membranaceous or transparent skin, extended between the tendons, or sinews; these are generally covered with small feathers of different shapes and colours, which present, when rubbed off, to the powers of the microscope, a most curious object, being placed over each other like the tiles of a house. When these are only inserted on certain parts of the wing, the rest appear transparent, similar to those of the common fly. The present moth is from a fine specimen in Mr. BULLOCK's collection, and is a male insect, the female is possessed of wings one inch larger in length and breadth; although this observation of the superior size is not always a certain criterion, as in some few species nature varies from that rule. The under wings are ornamented by a beautiful chain-border, the upper wings consist of red and black streaks at each extremity, the space chiefly occupied with undulated dashes of black and brown, the antennæ are short and plain, and the whole presents to the view the true character of the Simplex Munditiis of Horace, which, if it could possibly be translated, might be denominated, elegance joined with neatness.

[This page was blank in the original]

Drawn by G.Perry.   Engrav'd by T.L.Busby.

STROMBUS SOLITARIS.

Pub.d by J.Stratford Jan.1.1811.

PLATE 52

[366]

CONCHOLOGY.

Genus---STROMBUS.
Species---STROMBUS SOLITARIS.

Character.—Shell univalve, spiral, the cheek or Maxilla
Oris turned backwards, and spread out like a flap,
the cheek also cut open and furrowed at the top
and bottom, near to where it joins the beak; the
beak twisted inwards and backwards.

AMONGST the various specimens of Shells presented
to our view in the marine kingdom, there are few which
possess a greater variety than the Genus Strombus, The
body is generally large, the spire and beak small, but the
most striking and characteristic mark is in the cheek of the
shell, which is very much expanded in the side, and
differs in that respect from all other Genera. A change
takes place during the growth of this shell to its full size,
which is extremely curious, and which is caused by the
operation of the animal which resides in it. As soon as it
finds its own body grow too large for the habitation which
it occupies, it uses a flexible instrument of the form of a
trowel, and with which nature has wisely provided it, and
spreads a natural viscous incrustation or plaister round the
edge of the mouth, gradually enlarging and adding to it,
as it progressively hardens; during this period it also
provides a safe covering for the instrument itself, which is
the spire or long horn that is represented at the the top of
the shell in the annexed engraving. Most of the other
Strombi however have five or six imitations of this principal
spire, and are generally placed upon the side of the cheek,
at regular distances. The inside of the Strombus Solitaris
(so named from the circumstance of its having only one

D d

horn or spine) is of a bright red colour, and the whole shell both within and without, is undulated and tuberculated, the beak and spire are sharp and pointed, and like the body adorned with red streaks. Its form is strongly contrasted with the Cypræa, for in the latter shell the flap of the cheek is always doubled and bent inwards upon the body.

The present shell which we have now executed is drawn from an original in Mr. BULLOCK's Museum, and is a native of Africa and the East Indies. It has more of singularity perhaps than beauty of colour, and resembles in its outline an Urn or Vase upon the side where the spire forms the principal termination. About fifteen species are at present known, of which the Strombus Chiragra is the most distinguishable for the great length and curvature of the spines, and smallness of the body; which makes it resemble the common star-fish so frequently found upon the English Coast. This curious shell with others of the Strombus Genus we purpose to insert in some future number of the ARCANA, hoping that they will form an interesting and comparative series of this singular division, to each enquiring Conchologist.

GENERAL REMARKS

ON THE

FORMS AND ANALOGY OF THE TOUCAN, PARROT, AND EAGLE.

BY MR. G. PERRY.

THE curious, yet sometimes almost insensible difference in the bill and feet, which exists in the divisions of the claws, has frequently suggested to my mind, the wish for a more exact analogy and description. I have therefore endeavoured in the account of three particular genera, to mark out some outlines of character, in a more distinct manner than has hitherto been done.

The Toucan is a native of Africa and the West Indies, and is also found in all the countries bordering upon the Torrid Zone. The bill of this curious bird is of an uncommon shape, as to its immense length and thickness, when compared with its body, which is sometimes rather small; the bill resembles the claw of a large Lobster, and this extraordinary bill is in one species seven inches and a half long, and flattened like the handle of a knife. The double billed Toucan, or Hornbill, has a second bill standing upon the top of the other, but something less in its size and length, yet it adds very much to the usual thickness of the head, and gives it a very heavy appearance. The generic name for these birds is Ramphostos, under this family we also include the Buceros of LINNÆUS, not seeing any sufficient difference for separating it. These birds, although so formidable in their appearance, are quite harmless and gentle; they feed principally on pepper, which they devour with great ardour, gorging themselves in such a manner as to void it crude and undigested, this however is no objection to the natives using it again; they even prefer it to that which is got fresh from the tree, and seem persuaded that the strength and heat of the pepper is qualified by the bird, and that all its noxious qualities are thus extenuated.

We also add that their bills are bent downwards, scythe-shaped, the upper mandible longest and bent over the lower one at the tip of the bill, although this is in a small degree.

The feet of the Toucan resemble those of a Parrot in in every respect, having on each foot two claws placed before and two behind, all of unequal length, and roundly hooked at the ends, the outside longest and thickest.

OF THE PARROT.

The bill of this bird so eminently distinguishing it from all other birds, is much shorter and rounder than that of the Toucan, and the upper mandible hangs out much farther over the end of the lower one, and is more hooked at the point.

The various circumstances of its shape are worthy of a close investigation. First, the bill quarterly formed, each part of the bill making an exact quarter of a circle. Secondly, the quarterly formed, projecting. Thirdly, aquiline: Fourthly, aquiline projecting; and Fifthly, by a crenated or channeled front. These distinctions seem to form the most striking varieties that we have hitherto seen, and are each of them to be found in the different Genera of former Authors, such as the Macaw, Parrot, Parroquet, Lory, &c. The Macaw has a circular rim, bare of feathers round the eye, which circumstance assimilates it to the Toucan, but separates it from the Parrot and other smaller birds.

For the Character of the feet in the Parrot we refer our reader to the foregoing account of the Toucan, they being exactly the same in all respects.

The Parroquettes seem on the other hand always to have a cheek and face covered, and are in some instances crested with tufts, spread like the larger Parrots, but the best division of these into sub-genera, would perhaps be from the form of the tail, which is a more sure test than the colour of the body or wings, which frequently vary in the male and female. The colours in the crest of the Hawk Parrot exhibit all the richest tints of the rainbow, and the bill is singularly marked with a hollow channel in the front, and ending square at the bottom.

The Papuan Lory is distinguished by the great length of the tail, its sharp bill ending in a crooked point, seem strongly to distinguish it from the rest of its congeners, and to assimilate it in some to the elegant birds of Paradise. The eyes of many of these birds seem to project forward, and by this means to loose that natural projection, which a hollow cavity would have improved, as in the human head, where the eyes is quared by a projection of the forehead and temples, but in these birds the deficiency (if so it may be called) is amply supplied by the nistitating membrane, which a protecting skin or membrane, with which the eyes is immediately covered over on any approach of danger, and at other times seems to tubricate and moisten the surface, as often as is necessary. We cannot help remarking in this the singular bill of the Jabiru, the lower mandible being turned upwards, something in the manner of the Avoset and a few other curious species, equally curious in that of the spoonbill, which terminates in an increaved breadth at the tip or extremity of the bill.

The Rhynchops Nigra is also a singular bird, of the size of twelve inches, found on the shoals of the Islands in the West Indies, the undermandible of the bill being about two inches longer than the upper. It is said to skim along the surface of the sea, many leagues from Land alternately dipping its bill into the waves for the capture of its prey. The Loxia also has in general the upper and lower mandible of the bill crossing each other for the convenient purpose of breaking open the core of different sorts of nuts, for which end it is admirably adopted.

Birds are distinguished chiefly from other animals by the following singularities. In the circumstance of their anatomy, they may be described when flying, as a ship having wings for oars, a bill resembling a bowsprit, its breast bone the keel, the legs and feet seem to be the only part not employed, but these hang down and answer for ballast to keep the vessel steady, and lastly the tail is a rudder of the best kind, to steer its passage through the ambient air. The neck is made beautiful and soft in its texture by the interposition of silky hairs placed amongst the feathers, and the quills of the wings gradually increases in their size from the origin to the extremity, and are capable by different joints, of being folded up closely to the body and serves to keep it warm, but the most remarkable conformation of all is in their bill, which answers the purpose of mouth and nose.

The Golden Eagle has been generally esteemed the sovereign of the feathered tribe, the dignity and majesty of its form are strongly indicated by his large and muscular neck, his powerful talons and broad spreading wings made him an object of admiration in ancient times, and the Romans adopted its image as the great standard for their armies, and the emblem of supreme power.

If the size and strength are to be supposed to constitute the superior claim to the title of king of the birds, perhaps the Condor Vulture* might dispute the right even with the Eagle, as the extent of his wings far surpasses all other birds that are known. The bill of the Eagle has an entire resemblance to that of the Parrot and Vulture-tribes, except that the latter has a large fleshy protuberance branching forth from the root of the bill. It has three lengthened toes in front and one behind, which, like the bill, is always of a yellow colour in the female, and from whence he has taken the title of the Golden Eagle. The male, in this genus, has this singular peculiarity attached to it, of being much smaller than the female, and which is usual also in the Hawks and Owls; it is found frequently upon the coasts of England and Scotland.

The Eagle is a solitary bird, and brings forth three or four young at a time. It generally chooses some high rock near the sea-coast, where he sits for whole days watching for his prey, the lonely tyrant of the stormy waste.

The Sea Eagle is a large majestic bird, his bill is very strongly bent, his eye fierce and frowning, and his food consists principally of fish, or carcases of dead animals thrown upon the coast, the bill is wholly black, the legs and feet of a bright yellow, and slender in form.

The Black Eagle is the smallest of the English, and is only two feet three inches in length, from the tip of the bill to the extremity of the tail.

The White-Tailed Eagle is conceived to be the strongest and largest of this tribe, its bill is very broad, the hook of it pointed and projecting in the extremest part; the nostril is deep and plainly marked; his whole features exhibit a haughty ferocity and untameable cruelty. The tail is broad, spreading and white, the back feathers of a light and dark brown mixed, like most of its congeners.

---

* See No. 2. of the ARCANA.

Having thus given a comparative outline of the most remarkable circumstances of the bills and feet, in what we have described of the Linnæan genera, we cannot help noticing the curious bill of the Pelican,* which is provided with a large flexible pouch, in the under-side of which he can carry away half a bushel of dead fish.

The Flamingo owes all its singularity to the amazing length of its neck and legs, the bill is broadest in the middle and bending suddenly inwards, something like the head of a walking stick, and the lower mandible also, three times as thick as the upper one. This is the only instance that is known in nature of a bird having such a form. The Horned Screamer has a kind of sharp bill springing out of the shoulder of each wing and supposed to be extended for self-preservation, and from the center of the forehead proceeds a sharp and curvated bristle hanging forwards over the bill. Such are some of the most striking contrasts and exhibitions of variety, not demonstrated always through a whole family of birds, but most likely brought forward by the hand of the Creator, to excite our wonder and surprize at the greatness of all his designs.

* We shall shortly have the pleasure of presenting our Subscribers with a representation of the Pelican, from the drawing taken from a beautiful living specimen in the possession of Mr. Polito.

[This page was blank in the original]

Drawn and Engrav'd by T.L.Busby.

MOUNTAIN COW.

Pub.d by J. Stratford Feb.ry 1811

PLATE 53

ZOOLOGY.

Genus---ANTELOPA. *Species*---ANTELOPA MONTANA; or, MOUNTAIN COW.

Generic Character.---Animal having a smooth hairy skin, the hoofs of the feet cloven, the horns short, bent backwards, annular, having one or more rings placed upon the horns, tail of a middling length, between the Cow and Goat, residing chiefly in the hottest climates of the globe.

AMONGST the various gradations of animal creation which adorn the different regions of the earth, the Antelope forms one of the most extensive and interesting families, and which is not yet fully explored through the lengthened scale of existence. The tall and graceful character of their forms, their modest eye and glossy beauty, their light and agile motions, have formed a very pleasing subject for the admiration of the traveller and the themes of the oriental Poets. Equal beauty indeed is not dispensed over the whole, as the circumstance of bodily and muscular strength are more necessary to those Antelopes which reside in rocky or mountainous situations, that the difficulties of climbing or leaping from rock to rock might be more easily overcome; those who live in the extensive plains and vallies of India and Africa are more slender and have a greater agility of form, with a lighter construction, calculated for swiftness. These latter reside in the fertile patches of land which invest the borders of the Zaara, or great Desert of Africa, also in the southern regions adjoining to the Cape of Good Hope. About fifty different kinds have already been described and these are supposed not to be one half of what exist in this immense continent. Tibet also, the most fruitful, and

E c

most mountainous region of Asia, is reported to contain an amazing variety. Europe indeed being colder in its temperature and climate, only affords one species the Scythian, which resides in Hungary and Turkey. In America it appears that three kinds of these animals exist, which are all of them closely allied to the Antelope (except their having no horns) *viz.* the Lama, the Vicuna, and Guanaca, one of which has been already described in a former number of the ARCANA and to which we refer the reader.—The curious animal we are at present about to illustrate, so far as our enquiries reach, is quite undescribed, and has been wholly unknown hitherto in England, we shall therefore in the first place, give our reasons for placing it under the present arrangement, and then proceed to a more particular description of its form.

The character of these animals so nearly assimilated to the Deer and the Goat, has occasioned to most naturalists a difficulty in the nomination of the genus. The Goat, the Antelope and Deer have been classed by the eminent naturalist LINNÆUS under the general and distinctive name of Capra; regarding the Antelope and the Deer as being only a larger species of the Goat. We should be rather inclined however in the present instance, to follow the arrangement, which has been laid down by Professor Pallas, in his Description of the Antelope Leucophæ, which is an animal wholly of a blue colour and inhabiting the central and hottest parts of the immense continent of Africa. The horns of that creature are inclined backwards, and adorned with circular rings or ridges the same as in the animal which we are now about to describe. The distinguishing marks of this beautiful tribe of creatures and in which they differ both from the Goat and Deer, are principally these; their horns are annulated, having a ring round them, and at the same time are marked with longitudinal depressions or furrows running from the base to the point. There is also a curious deep channel or cavity placed beneath each eye. Besides the extreme beauty and meekness of its aspect, the Antelope is more delicately and finely limbed than the Roebuck; its hair is smoother and more glossy; its swiftness, which insures a complete degree of safety from the Greyhound, the fleetest of dogs, induces the sportsman to call in the aid of the Falcon, which being trained to the work, seizes on the animal or impedes its motion, so as to enable the Dogs to overtake it. In Persia and India a sort of Leopard is sometimes used in the chace, which takes its prey by the greatness of its springs, but should it fail in the first essay the game escapes.

Some kinds of the Antelope form large herds of two or three thousand, while others remain in small troops of five or six. They live for the most part in hilly countries, and browse on the tender shoots of young trees, which give their flesh an excellent flavour.

The Pasan Antelope described by Buffon is famous for a concretion in its stomach or intestines, called the oriential Bezour, which was much esteemed in former times for its great virtue in expelling poison from the human frame and was sold at enormous prices, its value increasing in proportion to its size. The virtues which ignorance and inexperience attributed to it, are now found no longer to exist and it is at present only consumed in countries where the knowledge of nature is but little advanced. Similar concretion are found in the bodies of goats, serpents, and hogs; in short there is scarce any carnivorous animal that does not produce singular substances in the intestines, stomach, kidnies and even the heart.

ZOOLOGY.

The Harness Antelope is remarkable for a collection of white lines, which cross the back in a square manner, forming a beautiful resemblance of a carriage harness, from which circumstance it has received its name. The body is of a rich and dark chesnut colour, with a very silky gloss of skin, the eyes are large, the horns bending backwards and slightly covered, the neck round, and like the body narrow, and elegantly formed. It seems as if Nature had intended this charming creature to bear away the palm of beauty from all the other species of Antelope. An excellent speciment is in the Museum of Mr. BULLOCK, and exhibits all the effect of animated nature, joined to the greatest harmony of colour.

The Mountain Antelope has a considerable resemblance to the Goat and is nearly of the same size with that animal, being about two feet six inches from the front of the breast to the tail, its general form is rather robust and muscular, its horns are short, bluntly pointed and marked near the bottom with a strong and thick ring, and are placed upon an elevated base. The face indicates much mildness, like the rest of its congeners, and the protuberance of the chest and shape of the hinder parts give it in some degree the contour of a Cow or Buffalo. It is a native of the mountainous parts of Morocco and is found chiefly on the hills which surround the foot of Mount Atlas, forming a part of a present of curious and interesting animals, lately sent over by the Dey of Algiers to his Majesty, and preserved alive in the Tower, having hitherto escaped from the frigid effects of a colder climate. It feeds principally on clover and hay, of which it is very fond, and in temper it is mild and grateful, attached to its keeper, and even to strangers who play with it or pass their hand smoothly over its head.

The Mountain Cow has a considerable hump upon the shoulder, and the neck is singularly shaped, bent in the manner of the camel; the tail is short and bushy as if cut off at the end.

The next Antelopes, perhaps, in point of size, which we shall shortly describe are the Memina and the Chevrotain or Dwarf, these are wonderfully small and delicate in their limbs, being only from twelve to fourteen inches in their height, when measured to the top of the body; the first is a native of Japan and the latter is found plentifully upon the Gold and Ivory Coasts, but its tender form is found to be wholly unable of bearing the rigours of any colder climate.

Such are some of the most remarkable species of the Antelope tribe, the present limits of the work will not allow us to enlarge more than by remarking that the Nylghaw of the East Indies, seems to be a direct inter-genus in its shape and character, between the Cow and the Antelope, and very difficult to decide, with which it should be classed, is of the size of a Stirk of three years old, and is much hunted after by the Indians for the value of the flesh. The reader will find it well figured and described by BUFFON in his excellent and extensive work upon quadrupeds.

CONCHOLOGY.

Genus---ARANEA.
Species---ARANEA CONSPICUA.

Character.—Shell univalve, spiral, the spire and body short and rounded, the beak long and armed with a triple row of spines, the mouth undulated and labiated. The body, spire, and beak, invested with a triple accumulation of curved and pointed integuments, open at the base.

IN a former number we presented to our readers, the representation of a Shell of the same family or genus as the present, under the name of the Aranea Gracilis or Slender Spider Shell. At first this might appear to the student in Conchology to be very different from the present one, which is denominated the Conspicua. Upon an examination however of the distinctive characters of each of them, it will be observed that the chief difference consists only in the number of spines which stand upon the body, beak, and spire of each, these being much more numerous in the Gracilis, the general form in other respects being very similar.

The native place of the present shell is wholly unknown, but it is supposed to be produced in the southern regions of the globe, it is drawn from Mr. BULLOCK's Collection which is of the same size, and forms a singular instance of a shell, of which we have not beheld the similitude of form in any other Museum which we have hitherto seen, it is very probably therefore an unique.

The character of the Aranea Conspicua consists of bold projections, the spines and turbercles are very few but

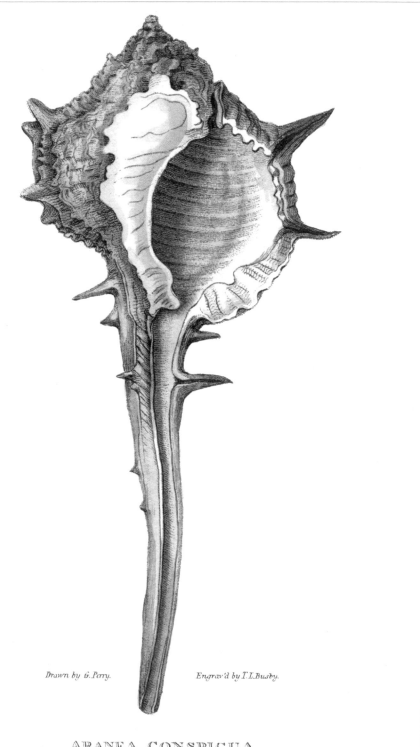

Drawn by G. Perry.　　　　　Engrav'd by T.L. Busby.

ARANEA CONSPICUA.

Pub.ᵈ by J. Stratford. Feb. 1. 1811.

PLATE 54

[This page was blank in the original]

strongly marked, the external colour of a pale brown, the inside of the mouth of a flesh colour, verging to a white, and the body round and closely covered with thickly placed turbercles.

Twelve species of this genus only are at present known, and they have certainly characters very distinct from the Murex, Buccinum, and Strombus, so as to be easily distinguished at first sight. Among these may be reckoned the spines, which are placed irregularly all over the shell and spring from the membranaceous divisions, which vary in number, in the different species, the mouth also is closely marginated by a projecting ridge, which almost shuts it out from the beak, the latter being much longer in this than in any other shell. The spines may perhaps serve as a guard or defence to the animal to preserve it from danger, and also to facilitate its rotary motion at the bottom of the sea. The beauty and variety of this genus, have always been strongly recommended to the notice of collectors, who think themselves very fortunate in obtaining perfect specimens of it.

Some of the Aranea Shell are of a blue colour, which is very singular, seldom happening. Nature indeed seems to have been very sparing in the formation of blue in the shell tribes, yet very profuse with it amongst the individuals, of the vegetable kingdom, the contrast however is delightful when there is only a small diffusion of blue, too much has a cold and unhappy effect, either in nature or the arts, by putting us mostly in mind of the rigours of winter, unless contrasted with brighter colours, such as the orange, red, or brown.

Genus---CONGIOPODUS.
Species---CONGIOPODUS PERCATUS.

Character.—Body concave, depressed in the middle of the back, the back fin membranaceous, armed with spines, erect and pointed; jugular fins two, one dorsal, and one ventral fin; tail divided, and the thorax projecting and long, sustaining one central fin, the head and nose angularly wrinkled and depressed.

AMONGST the various and singular tribes of fish which are hitherto unknown to Natural History, either by description or representation, and have eluded by their scarcity, the observation of the mariner or traveller, the Congiopodus may be reckoned one of the most remarkable in its form and character.—The body of this fish is rounded and of a dark olive colour, spotted with irregular oblong marks of a deep brown, the forehead is large and hollowed out, the nose truncated, or cut off at the end: one may observe in this fish a general resemblance to the Dolphin; but the Dolphin has the back membrane with its spines, continued as far as the tail, which is different in the present fish, being interrupted by a standing dorsal fin, and an open space finishing plain to the root of the tail.

The scales of the Congiopodus are either very small so as to be invisible without a microscope, or else has a gelatinous skin, which last is the most probable, and these two are the only modes that nature seems to adopt in the coverings of fishes. The scales are sometimes oval shaped, oblong, or notched, and the fins which in some species form a flexible membrane, resembling an arm or hand,

Drawn by G.Perry   Engrav'd by T.L.Busby.

CONGIOPODUS PERCATUS.

Pub.d by J.Stratford Feb.y.1.1811.

PLATE 55

[387]

[This page was blank in the original]

composed of different joints, are in other fish, nothing more than flat elastic bones standing out from the body, without motion or action. Such are the Diodon and others which exhibit sometimes a surface covered with prickley spines, which are no doubt intended for self-defence, or to annoy their enemies, others have these horns placed on their back like those of an animal. The Unicorn Diodon has one straight horn, sharp, tapering, and growing out of the middle of the forehead, the body is square when viewed sideways, but if seen from either end it appears triangular Thus the habits and propensities of fish are as various as their forms, whilst some by their various qualities are wisely destined to thin the over numerous swarms which abound in shallow coasts and rivers; others are singularly defended by a curious coat of armour, resembling the skin of a Rhinoceros, or by a !most deadly weapon fixed upon their beak, as the Swordfish and the Narwhale; most of the fish hitherto known, have one eye placed on each side, and these are often of the flat kind, as that which is called the Hallibut, the John Dory and the Stromateus, but other flat fish as the Plaise and Flounder, have their eyes placed on the same side, and the jaw is always unequally formed and placed obliquely. The names of the fins which may not improperly be called their arms, have been distinguished generally in the following manner. The dorsal or back fins, the ventral or belly fins, the anal or vent fins, and the caudal or tail fins, and lastly, the cartilaginous fishes, have only a membranceous skeleton, instead of a boney one, as their names imports.

F f

Genus---PELICANUS.
Species---PELICANUS AFRICANUS.

Generic Character.—Legs short, feet webbed, and spread
out into three angular divisions, the body oval,
the neck very long, body square at each end and
covered with flowing loose feathers, which disclose
underneath in certain parts, the larger quills.
The head is crested with the same feathers, the
bill long, acuminated, and bent down at the end.
A large elastic pouch for holding food is placed
on the mandible.

AMONGST the various provisions which nature has
imparted for the preservation of the different species of
birds, few are more remarkable than the pouch with which
the Pelican is provided, and this, when distended into
an enormously large bag, by being filled with fish, resembles a basket lined with leather. With this load, it is said
frequently to fly to an immense distance, and lessening the
pressure of the weight upon the neck, by resting the middle
of it upon its breast, it alters the centre of gravity, by
which circumstance, the motion necessary to its passage
through the air, is more compleatly balanced.

The Pelican of the Wilderness, for by that name it
has been known by ancient writers, (from its solitary life
passed in desart places) has been fabulously reported to
feed its young with its own blood, and artists, in painting,
who generally adopt the opinion of the poets, have repeated
the same error in all their representations of this bird.
There may be some plausible reason indeed in the natural

Drawn and Engrav'd by T.L.Busby.

PELICAN.

Pub.d by J.Stratford Feb.y 1.1811.

PLATE 56

[391]

[This page was blank in the original]

appearance for the production of such a mistake, since in the extensive and arid plains of Africa and Asia, where the pools or rivers are dried up by the heat of the climate, food of course is scarce, it must be collected in quantities ; the parent bird no doubt adopts this expedient for supplying its young with nourishment from distant regions; thus giving the idea of their feeding on the mother. There is a very considerable resemblance in the Pelican to the Dodo, a bird of the duck kind, found in the East Indies, but which has a more ornamented character in its colours and head than the former. The colour of the Pelican is of a pale pink shaded in certain parts with a darker tint of the same colour, the bill is yellow and remarkably strong in the formation of its membranes, which are capable of very great extension ; the legs and feet are thick and curiously turned inwards with short claws. The curvature at the end of the bill seems to be well calculated for breaking or tearing its food, and it is irregularly bent in certain parts to keep its prey from escaping out, when first it has seized it in the water. The Physiognomy of its eyes and visage is striking, its walk is slow and solemn ; the long arched neck, like that of the swan, is capable of seizing any object that is below or round him ; on each side of the head is placed a white angular mark which is without feathers, which encloses the eye compleatly. The Pelican is capable of being entirely tamed, and of uttering certain loud sounds or screams, which are not disagreable to the ear, and he seems to form an inter-genus of the duck, the swan, and the crane. The general length of his body and tail, is about three feet.

The present noble specimen is in the possession of Mr. POLITO, Exeter 'Change, and forms a very animated instance of this rare bird.

Remarks on the Locusta, or Gryllas, a native of Arabia Felix.

SYRIA, Egypt, Persia, and almost all the south of Asia, are subject to a calamity as dreadful as volcanoes and earthquakes to other countries, in being ravaged by those clouds of Locusts which produce the famines so often mentioned by travellers. The quantity of these insects is incredible to those who have not themselves witnessed their astonishing numbers; the whole country becomes covered with them for the space of several leagues. The noise they make in browzing on the trees and herbage may be heard at a great distance, and resembles that of an army foraging in secret. The Tartars themselves are not a more destructive enemy than these animals. One would imagine wherever they have been seen, that fire had followed their progress. Wherever their myriads are spread out, the verdure of the country disappears, as if a curtain had been removed; trees and plants are stripped of leaves, and reduced to their naked boughs and stems, so that the dreary images of winter succeed in an instant to the rich scenery of the spring. When these clouds of Locusts take their flight, to surmount any obstacle, or to traverse more rapidly a desart soil, the heavens may be said literally to be obscured by them. Happily this calamity is not repeated frequently as it is the inevitable forerunner of a grievous famine, and all the maladies which it occasions. The inhabitants of Syria have always remarked, that Locusts are increased by winters too mild, and that they constantly come from the desert of Arabia. From this observation it is easy to conceive that the cold not having been rigorous to destroy their eggs, they multiply suddenly, and the herbage failing in the desert, innumerable legions issue forth. When they make their first appearance on the frontiers of the cultivated country, the natives attempt to drive them off, by raising large clouds of smoke, but frequently the smell of burnt herbs and wet straw fails them. They then dig trenches, where numbers of the insects are buried, but the most efficacious destroyers are the south and south-easterly winds and the locust-eating thrushes. These birds follow them in numbers like starlings, and not only greedily devour them, but kill as many as they can, accordingly they are much respected by the peasant, and no one is allowed to shoot them. By the south and south-easterly winds, the clouds of Locusts are driven over the Mediterranean, where such quantities of them are sometimes drowned that when their carcases are thrown on the shore, they infect the air for several days, even to a great distance. Locusts, it is said, have sometimes been observed in Britain, and great mischief has been apprehended; but the coldness of the climate, and the humidity of the soil are very unfavourable to their production; they therefore all perish, without leaving a young generation to succeed them.

We are informed by different travellers, that the touch of the Locust is almost as injurious to vegetables as what they devour, if so their bite must contaminate the plants, and either greatly corrupt or destroy their vegetation: the female Locust generally lays about forty eggs in some solitary place, inserting them in the ground. It is strange that so horrid an insect should be esteemed excellent food, but tastes alway differ, and the Negroes of Gambia eat them, broiled on coals, or pounded and boiled in milk.

REMARKS ON THE COCHINEAL INSECT.
(of South America.)

THIS substance which is found so useful in commerce, for the purpose of painters and dyers, is nothing more than a dried insect, which is found upon a particular plant, called the Cactus Opuntia or Prickley Peartree. In

Jamaica and also in some parts of America, these insects are pretty common. When the young brood called the Cochineal insect are arrived at their full growth, they adhere to the cactus in a torpid state; it is in this state that they are taken from the plant for use. Twice or thrice a week, the slaves appointed to this employment go among the plants, and pick off carefully with a bamboo twig shaped somewhat into the form of a pen, every full grown insect they can find; a sufficient number is left for the production of future broods. The insect are converted in the substance called Cochineal, by the following process; the insects which are collected in a wooden bowl, are quickly spread from thence upon a thin dish of earthen ware, and placed alive over a charcoal fire, where they are slowly roasted, till the downy covering disappears, and the aqueous juices of the animal are entirely dried up, and a little water is carefully sprinkled on the parts to prevent burning, which would entirely spoil them: they then appear like small dark reddish grains and take the name of Cochineal, preserving so little of the original form of the insect, that this most precious scarlet dye was long known in Europe, before naturalists could determine whether it was an animal, vegetable, or mineral substance.

The male and female, when alive, have each six legs, the female has no wings like the male, which has two, erect, and of a straw colour, the male is of a scarlet, and the female of a pink colour similar to lake. It has been computed that there are imported into Europe, in the course of trade, no less than eight hundred and eighty pounds weight of cochineal, annually.

OF THE LAC COCHINEAL INSECT
(of the East Indies.)

THE head and trunk of this insect seem to form one uniform, oval, and compressed red body, somewhat of the shape and size of a very small head of a pin, consisting of twelve rings. The Antennæ are half the length of the body, filiform and diverging, sending off two or three diverging hairs, the tail is a little white point with two diverging hairs. The male and female have not been yet distinguished perhaps for want of proper glasses. The eggs are brought forth in November and December, for some time they traverse the branches of the trees upon which they are produced and then fix themselves on the extremities of the young shoots. By the middle of January they are all fixed in their proper situations and now exhibit no signs of life, appear as plump as before. The limbs, antennæ and bristles of the tail surrounding them in a beautiful red liquid. These insects which in the East Indies have the name of Gum Lac are principally found on the trees of the uncultivated mountains on both sides of the Ganges, where nature has been so bountiful, that were the consumption many times greater than it now is, the markets would be fully supplied. The best Lac is of a deep red colour, if it be pale and pierced at the top the value diminishes, because the insects have left their cells and consequently it cannot be used for dying, though probably it may be of more value as a varnish. This substance is now principally confined to the making of sealing wax, and to japanning, painting and dyeing.

OF THE WAX-FORMING CICADA
(of America and China.)

THIS is a singular Insect, and deserving of some attention, both as an object of curiosity, and from its importance in domestic œconomy. Its winged cases are green, margined with red and deflexed; the interior ones are spotted with black. In the variety, figured and described by SIR GEORGE STAUNTON, these are whitish, margined with black, and have a row of black spots on the posterior edge. The Larvæ are elegant and beautiful creatures and to their labours the

Chinese are indebted for the fine white wax that is so much esteemed in the East Indies. They form a sort of white grease, which attaches to the branches of trees, hardens there and becomes wax. This is scraped off in the autumn, melted on the fire and strained, it is then poured into cold water, where it coagulates and forms into cakes. In appearance, it is white and glossy, and mixed with oil, is used to make candles, for which purpose it is thought greatly superior to bees-wax.

The Insects are white when young, and it is then that they make their wax; when old they are of a blackish, chesnut colour and fasten themselves to the external branches of the trees. These at first are each of the size of a grain of millet, but towards the beginning of the spring, they increase in bulk and spread; they are attached to the branches like grapes and at first sight the trees which bear them appear loaded with fruit. About the beginning of May the inhabitants gather them and having them in the leaves enveloped and inclosed by a broad-leaved grass, suspend them to the trees; at the end of July they form their wax.

Sir George Staunton says of these insects, that he saw them busily employed upon the small branches of a shrub, that in its general habits, had a strong resemblance to privet. They did not much exceed the size of a domestic fly and were of a very singular structure, being every where covered over with a white powder and the branches they most frequented, were entirely whitened by this substance which was strewed all over them.

> " Happy creature! what below
> Can more happy live than thou?
> Seated on thy leafy throne
> Summer weaves thy silver crown."

[This page was blank in the original]

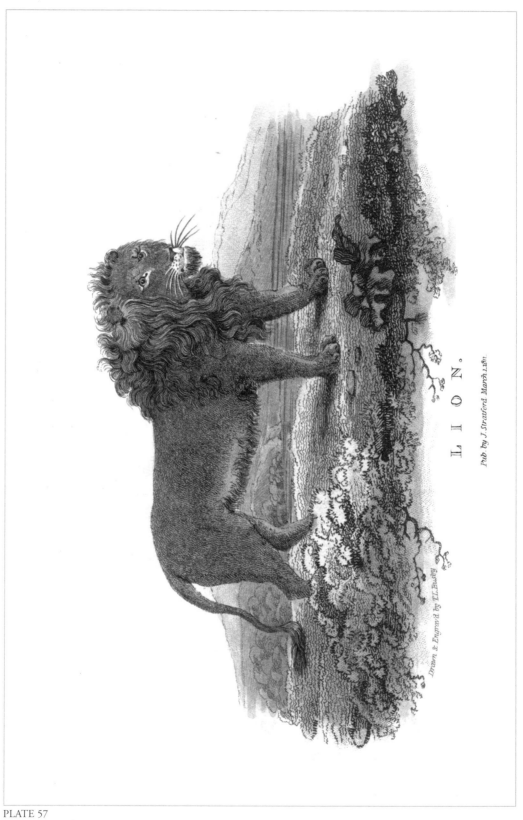

L I O N.

Pub. by J. Stratford. March. 1.1811.

Drawn & Engrav'd by T.L.Busby.

PLATE 57

ZOOLOGY.

Genus—FELIS LEO. THE LION,

Generic Character—Six teeth, four central ones equal in
length, three grinders on each side; the claws
extending and retractile, the tongue armed with
prickles bent backwards; the neck covered with
a long and hairy mane.

AMONGST the numerous tribes of destructive animals,
which infest the solitary regions of the torrid zone, the Lion
stands pre-eminently conspicuous. His look at once con-
firms him the chief of quadrupeds, whether we consider the
majesty and strength which his muscular forms indicates,
or the solemn march of his deportment. His steps are
more firm and commanding, his looks more generous and
noble than those of any other creature; he mostly refrains
from any attack upon the human race, unless highly incen-
sed by his foes, and he proudly scorns the smaller animals
as a mean and ignoble prey.

The Lion is found in the warmer regions of Asia, but
attains his highest perfection in the interior of Africa. His
strength is such, that with a single stroke of his paw he can
break the back of a horse, and he has been known, not unfrequently to carry off a young buffalo in his teeth. He
rarely engages in full day-light in the pursuit of his prey,
but on the approach of night, quits his habitation and with
a hollow roar, which over-whelms all the other creatures of
the forest with consternation, commences his career of
havock and destruction. It is fortunate that his sense of
smelling is not very acute, for he depends in the chase only
upon actual sight or probable inference, he frequently con-
sumes at one repast sufficient food to satisfy him for three

g G

or four days. He breaks the bones of the buffalo with perfect ease, and frequently swallows them; the reversed prickles of his tongue are of extraordinary strength and extension, and serve as a rasp to draw the blood. After a full repast, he returns to his den, and enjoys a state of slumber, and repose, until the calls of hunger rouse him again to fresh activity: the Lion in full exertion of his energy, must present to the mind a most impressive image; the majesty of his countenance, set off by his full flowing mane, his eyes glaring with indignation, and all the apparatus of destruction in his mouth and paws, he must form sublime and admirable materials for terrific imagery.

At the Cape of Good Hope, it is not uncommon to hunt the Lion in an open and spacious plain, in which he finds it impossible to escape his pursuers by flight, he checks his progress and boldly fronts his adversaries, awaiting their attack: several dogs which first dare to attack him, generally fall under his stroke, but in a few moments he is over-whelmed by numbers, and literally torn to pieces. The Negroes of the Cape, are reported to eat his flesh, and his skin which was formerly a mantle for a hero, is now more frequently imployed for the bed of a Hottentot.

The Lion has been known to measure eight feet in length, exclusive of the tail which is about three or four. Its colour is of a pale tawny, and the male possesses an extremely full and flowing mane; the female is destitute of this, and is considerably smaller than the male. It has been known to live in a state of confinement, to the age of sixty-three, or even seventy years; though from a philosophical examination of its general structure, it would be concluded only to exist for half that period. The female only produces one litter consisting of four or five in the year, which at first are extremely small, scarcely exceeding the size of a half grown kitten; they are six years in attaining their full growth.

It is imagined that Lions are much less numerous in Africa now than formerly, as a very large number were formerly exported to Rome every year. MARK ANTHONY is said to have been drawn through the streets by tame Lions harnessed to his chariot, to furnish entertainment for the inhabitants of that splendid and luxurious city. Lions were conveyed in vast numbers from the interior of Africa, to exhibit at the public festivals, where they might fight with each other, also with other animals, and even at length with men; these diversions were first exhibited by QUINTUS SCÆVOLA, but were afterwards carried to a far greater extent by SYLLA, who displayed in the arena a hundred Lions during his prætorship. JULIUS CÆSAR to conciliate the people, entertained them with no fewer than four hundred; and POMPEY imported at a vast expence and elaborate research, upwards of three hundred at one time. In proportion as population has been extended, and national intercourse advanced, their range has become more limited, and their acquaintance with man seems to have considerably checked that daring, which was supposed by many incapable of being daunted. The Lion's valour diminishes in proportion as he resides near the habitations of men, whose ingenuity and resources he seems well aware must always secure them a superiority in the conflict with other animals. In the neighbourhood of the small towns of Africa, even women and children have not unfrequently driven Lions from their lurking places. When taken young they can be taught to sustain confinement without difficulty, and will not only shew tranquility and contentment, but occasionally engage in sports and gambols with smaller animals, with which they have been gradually associated. They are susceptible of attachment and gratitude, will caress their keepers, diplay a magnanimous forbearance with respect to the offensive freedom and petulant insults of weaker creatures, and after having once, as it were, pledged themselves for its security, for any one which by an act of wantoness may have been thrown into its den, will endure extreme hunger before it can permit itself to destroy them. The natural excitability of these animals however is so great, that all the discipline of education is frequently insufficient effectually to repress their passions within secure bounds ; and in some unlucky coincidence of circumstances, those familiarities with them, which had been permitted without the slightest inconvenience or reluctance, have proved fatal to the persons who engaged in them : although the Lion frequently attacks his prey in open chase, he generally adopts the system of ambuscade, and will lurk on his belly in some thicket, frequently near the water, awaiting the approach of any animal which its evil destiny may impell near it, on which he will spring with a sudden bound, rarely failing of success, and sometimes reaching to the distance of twenty feet ; when the leap is unsuccessful, the object is often permitted to escape without pursuit, and he retraces his steps slowly to the thicket, as it were abashed by his failure, and anticipating consequences of greater adroitness in the ensuing efforts.

The present representation is delineated from a living one, which has nearly compleated its full growth, and with many curiosities is now preserved in Exeter Change, London, and may be justly considered as the most perfectly beautiful specimen ever met with for many years in the metropolis. It appears also that there is a curious variety of the Lion found in Barbary of a much less size than the above, and having a much smaller mane.

[This page was blank in the original]

Drawn by G. Perry.   Engrav'd by T L Busby.

BUCCINUM DILATUM.

Pub. by J. Stratford March 1.1811.

PLATE 58

[406]

Genus—BUCCINUM.
Species—BUCCINUM ORBICULARE.

Generic Character.—The shell spiral, ovate or elongate; the cheek standing straight forwards, the opening oblong. The beak invested at the back with a thick, twisted collar, swelling outwards, no canal; the body covered with tubercles and continuous bands.

THE genus Buccinum has too often been confused by the early writers upon Conchology, with the murex, harpa, dolium, and cassis; the distinction however, is very plain and obvious, by means of the thickened collar which invests the beak, and forms a small umbilicus, or hollow navel underneath; ending in a curved and twisted form. The spire also, is in general rather short, and rounded at the summit; the body and cheek of the mouth bulbous and projecting; the whole shell is covered with circular lateral ribs, having a multitude of large and small red spots, upon a ground of a pale red, verging to a brown. There is an amazing lightness and elegance in the general appearance of this specimen, and the stripes of the internal surface are finely contrasted in direction as well as colour, from the outer surface, which terminates at the base in a smooth pointed beak or channel. The fish which inhabits the Buccinum shell, does not enlarge its mouth during the period of its growth, by any extraneous addition of calcareous matter, as is the case with the cyprœa, the strombus, and others, but retains the same character of form and shape, without any obvious variation. The present species i found at the Cape of Good Hope, and adds a pleasing and ornamental appearance to a collection of shells, when contrasted with darker colours.

CONCHOLOGY.

The Buccinum is sometimes found in England in a fossil state, along with the oyster and cornuammonis, in Derby-shire, Hampshire, and almost all the other counties of England, they are discovered most plentifully imbedded in strata of clay or gravel. These species however, are much smaller than the present Buccinum, and present a curious circumstance to the naturalist, in having the mouth and spire revolving the opposite way, viz. from left to right, and excepting this difference, they have in other respects a strong resemblance to the common whelk, which is found so abundantly upon the English coast. From this cause they have sometimes been denominated by learn-ed authors, the heterostophe shells, and it happens, not unfrequently, to the helix and columna genera. Along with these, is often found the pholas, embedded in solid rocks of quarry-stone, or in modules of flint. This very singular fish inhabits a bivalve shell, resembling in some degree that of a muscle, only rough externally. It has the power, it should seem, of penetrating, by means of its muscular tongue, into the hardest rocks, dissolving them at the same time, as it is supposed, by a peculiar fluid, but the cavity of its cell or habitation being larger at the inner end than at the mouth, the creature as it grows larger becomes incapable of returning back, and perishes in a sepulchre of its own forming. The seeming unfitness, however, of this animal for penetrating into rocks, and there forming a habitation, has induced many writers to suppose, that they entered the stone while yet in a soft state, and that some petrifying quality had hardened the whole substance by degrees. This latter opinion however, has been ably confuted by Dr. Boads, who discovered that the ancient pillars of the temple of Serapis at Puteoli, were penetrated by them after it was erected, as such stones would be unfit for the original erection of a temple.

BLACK SWAN.

Drawn by I. C. Whichelo

Engrav'd by I. I. Busby

Pub. by I. Stratford, March 1. 1811.

PLATE 59

ORNITHOLOGY.

Natural Order-ANSERES. Genus-ANAS CYGNUS. Species—NIGER.

Generic Character—Bill strong, broad, depressed, and furnished at the end with a curved point, the edges of the mandibles serrated, with sharp teeth, nostrils small and oval, toes four, three before and two behind, the middle one the longest.

THE Genus Anas of Linnæus, includes a large family of differently formed birds from the swan downwards to the teal. Of these the wild and tame swan, though both natives of England, are very opposite in their forms, and supposed to be distinct species, since no instance has been found of their uniting in society ; and the latter weighs half as much again as the former, being twenty-five pounds weight, the former much less. The ancients were of opinion that no such bird existed as a Black Swan, and were apt to use the following proverb when any thing improbable or impossible was related, " Rara Avis in Terra simillima nigro Cygno," " a rare bird something like a Black Swan." Modern discoveries in natural history have produced from New Holland and Van Diemen's Land a Swan of the blackest kind hitherto known ; perhaps to convince the sceptical of many things in heaven and earth that are not dreamed of in our philosophy. Previously to the description of the Black Swan, we shall give a brief account of the two European species which are perfectly white. The Anas cygnus ferus, or Wild Swan, measures five feet in length, and above seven in breadth, and weighs from thirteen to sixteen pounds, the bill is three inches long, and of a yellowish white, from the base to the middle and thence to the tip black, the bare space above the eye and eye-lid, yellow, the whole plumage of the bird is of a full white. At the setting in of winter they assemble in immense quantities, particularly on the large rivers and lakes of the thinly inhabited northern parts of Europe, Asia, and America, but when the weather threatens to be too stormy, they remove to Britain or some warmer climate.

The second species is denominated by LINNÆUS, the Anas Mansuetus, or Tame Swan, and is so generally known as hardly to need any description, the bill and skin between the eyes are red, differing wholly from the former. The Swan, though endowed with great power and strength, yet molests none of the other water birds, but when employed in the cares of a young brood, it is not safe to approach them, for they will fly upon any stranger, whom they often beat to the ground by repeated blows; and they have been known by a stroke of the wing to break a man's leg. The present Black Swan however, has several feathers scattered in the breast, the under-quills of the wing and also of the tail, which are of a dingy white. The neck is much straighter than that of the common Swan, and the whole bird is considerably less. The bill is variegated by stripes of red and white, and the crown of the head is depressed in the middle in a singular manner. When provoked or assaulted, it issues a loud hissing noise similar to that of a goose, but does not erect or expand its wing when sailing along the stream so much as the European Swan. It is delineated from a live specimen in the collection of Mr. POLITO, and seems to bear the English climate with as much health and vivacity as its native country. Thus, in process of time, if found to be useful as an article of food, it may perhaps be joined to the rest of our domestic water fowls, in the same manner as the Canada and Morocco goose, adding a pleasing variety to the appearance of our ponds, rivers, and parks.

[This page was blank in the original]

Drawn by G. Perry.

Engraved by I.L.Busby.

PAPILIO CATENARIA.

Pub.d by J. Stratford, March 1.1811.

PLATE 60

Distinctive Genus—PAPILIO.
Division—ARCUATUS. ARCUATUS CATENARIUS.

Character—The body covered with hair, the wings plain, the antennæ armed with a capitulum or small knob placed at each extremity.

THE present elegant fly is brought from the Brazils, and has a considerable resemblance in its form to the Cæruleus, before described in a former number of this work. There is this difference however, that the one now before the reader is drawn from the underside, in order to shew the beautiful markings which are not so perceptible when viewed from the back; the legs also are visible in this case, which are always hid by the wings in the other,

The upper wings of the Catenarius are ornamented with a rich profusion of ornaments, placed at the edge resembling gothic work or pointed windows, and having three black spots, annularly formed, in each upper wing. The center of these is also adorned with two strong shoulders or straps ending in various irregular lines, and which answer in some degree to the marks of the wings below. The under wings present to the eye a curious piece of chain-work, suspended from each side, and ornamented with beads irregularly indented and joined together at the lower end. The colour of the ground of the wings is pale blue, edged with white; the border on the outside and principal tendons are of a warm brown colour. Three rows of rich interlaced gothic arches surround the edges of the lower wings; these are variously formed, some angular, some of different size, placed with an irregularity (if we can be allowed such an

H h

expression) similar to the rich tracery of a cathedral, or the grand embroidery of the frost, representing a silver assemblage of lances, branches, and leaves, upon the surface of the windows in December.

The English Butterflies are far exceeded in size and magnificence by the equatorial insects, these latter being often four times as broad and large, but true elegance does not depend upon relative size, since the microscope offers to a minute inspection many of the smaller Moths and Papilios of England, decorated with the richest and most various metallic lustres that the mind can possibly conceive, and which the large ones are entirely without. Where the noble pine lifts its spreading and lofty branches to the clouds, such large creatures as the Atlas Moth and others will be found to be duly proportionate; and every division of animated nature seems agreeable to the sphere and scale upon which it is to act. How admirable is the balance by which the whole creation is kept up, and yet diminished, the different classes which fill the earth, air, and sea, with an innumerable progeny would sweep over and cover the whole habitable surface, were it not for the destructive powers of the Vulture, the Eagle, the Lion, the Serpent, and lastly the devouring Shark, which swallows up myriads. We naturally imagine the insect-destroyers to be sent for the same purpose; the Ichneumon fly, the Aphis, Ant, &c. have no doubt the power of diminishing, in all their stages, the Butterflies, Moths, and innumerable other insects, which lie dormant for the colder part of the year, and only appear in the bright and sunny period of summer, to search for their usual and annual food, and then resign themselves a prey to their more voracious neighbours.——

"The golden chain of being melts away!
And the mind's eye sees nothing but the present."

Remarks on the different Genera of Insects as elucidated in the Seven Orders of LINNÆUS.

THE insect division of the world, first received its name from the general form of the individuals having a separation in the middle of their bodies, by which they seem to be cut into two parts; these parts are generally connected by a slender ligament or thread.

Insects are for the most part furnished with wings, as well as scales or rings covering the surface of their bodies, but capable of being easily moved by muscular bands; the mouth is in general situated under the head, and is furnished with transverse jaws, lips, tongue, teeth, and palate; it has also four or six palpi or feelers: insects have also moveable Antennæ proceeding generally from the front of the mouth, and which are endowed with a very nice sense of preception. From the situation of an egg in which they first appear, placed in some warm residence by the provident care of the parent, they undergo a metamorphosis, or change in three different ways and periods. First, from the egg to the Caterpillar or Larva; secondly, into chrysalis which it spins itself; and thirdly, into the fly or perfect insect, soon after which it lays a fresh brood of succeeding eggs, similar to the first situation.

The duration of these minute creatures in their most perfect state, are in general short, that the parents have but seldom an opportunity of seeing their living offspring. Of course they are not provided with milk-like viviparous animals, nor are they like birds, impelled to sit upon their eggs in order to bring their young to perfection. In place of these, the Supreme Power has endowed each species with the astonishing faculty of being able to discover what substance is fitted to afford the best food for its young, although many of these substances are so totally opposite to what the parent would eat, as no doubt to prove a poison to it in its grown state. Some of them attach their eggs to the bark, or insert them into the leaves of trees or other vegetables; some of them form nests which they store with insects or caterpillars, that will attain the exact state in which they will be proper food for their young when they wake into life, others bury them in the bodies of insects, or fall upon astonishing contrivances to convey their offspring into the dead bodies of other animals. Some insects drop their eggs into the water, as if they foresaw that such an element was necessary to their progeny in the first state of existence, although they themselves would soon be destroyed by it: in short their different modes of disposing of their young, seem to be as opposite as their various shapes, and certainly beyond all human enumeration.

LINNÆUS has divided the animals of the insect kind, into seven orders, viz.

1. Coleoptera (a sheath and wing) derived from the Greek.
2. Hemiptera (a half-wing) from Ditto.
3. Lepidoptera (a scaley-wing) from Ditto.
4. Neuroptera (a net-shaped or nervous wing) from Ditto.
5. Hymenoptera (bee-like wing) from Ditto or a membrane.
6. Diptera (double-wings) i. e. two standing together from Ditto.
7. Aptera (without a wing) from Ditto.

The Coleoptera Fusca, or brown beetle, denominated the melolontha of LINNÆUS, exists in such abundance in

Great Britain, that they hardly require from us any **very** minute description, they particularly frequent the sycamore, the lime, and the beech trees. The eggs are deposited in the earth by the parent, and sometimes remain there in **the** larva state for three or four years; they are sometimes known by the name of the cockchafer or may bug. Of superior beauty to this, but not quite so large, is the English rosechafer or green beetle, the auratus of Linnæus, which is of a burnished gold colour shaded with a purple and green, according to the way in which it is held. They are to be found on flowers, most commonly at the roots of trees, and never appear on the surface of the ground unless disturbed by digging, or some other accident. They are thought to be injurious to the gardener, from their devouring the roots of his plants and trees. The female deposits her eggs in the middle of June: for this purpose she burrows in the ground where it is soft and light, hollowing out and forming for them a proper receptacle; when the operation is over she flies off, but seldom lives more than two months after. The grubs are produced in fourteen days, and immediately seek out for food, which the parent always takes care to have near the place where she lays her eggs; as soon as they have attained sufficient strength, the young grubs separate each burrowing a different way in search of roots. They remain four years in this state annually changing their skin, till they become of full growth, when they are of a cream colour, with brown head and feet. During the winter they eat but little, if at all, and retire so deep into the ground, as to avoid the effects of the frost.

About the month of March, at the end of the fourth year, the grub forms a case of earth, about the size of **a** walnut somewhere near the surface, where it changes into **a** chrysalis, and in May bursts out as a perfect chafer.

The Lampyris noctiluca, or glow worm, is distinguished from some other of the Coleoplera tribe, by having the antennæ thread shaped. The thorax is plain, slightly orbicular and conceals the head. The segments of the abdomen is terminated in folded papillæ, and the female in most of the species is destitute of wings. In the English Glow Worm each sex is luminous, but in the male the light is less brilliant and confined to four points, two of which are situated on the two last rings of the abdomen.

> " Among the crooked lanes in every hedge,
> The Glow-worm lights his gem, and thro' the dark
> A moving radiance twinkles."

In the month of June this elegant speck of Nature seems to be the most plentiful, and its use is said to be much the same as that which was suspended for LEANDER, to light him across the Hellespont. It has been attempted to account for this curious phœnomena in the following manner; it is probable that the phosphoric acid is produced by a voluntary combination of oxygen gas, and that a light is given out through their transparent body by this slow internal combustion. By contraction, the insect can instantly withdraw the light, which is perhaps rather an argument against any chemical decomposition.

The Cerambyx Violacens or Timber Beetle, is of a dark violet colour, rather hairy and punctured; the wing cases are narrow, rounded at the tip and bulking towards the base. This insect both in the larva and perfect state, feeds principally on fir timber which has been felled some time without having the bark stripped off; but it is often found on other wood: though now too common in this kingdom it is supposed not to have been originally a native.

The circumstance of this destructive little animal attacking only such timber as has not been stripped of its bark, ought to be attended to by all persons who have any concern in this article, for the bark is a temptation not only to this but to various other insects, and much of the injury done in timber might be prevented, if the trees were all barked as soon as felled. The female is furnished with a flat retractile tube which she inserts into the wood about a quarter of an inch deep, and there deposits a single egg.

By stripping off the bark, it is easy to trace the whole progress of the Larva, from the spot where it is hatched to that where it attains its full size. It first proceeds in a serpentine direction, filling up the cavity with a kind of dust, which stops all ingress of its enemies from without: when it has arrived at its utmost dimension, it does not confine itself to one direction but works in a kind of labyrinth, eating backwards and forwards, which gives the wood under the bark, a very irregular surface; and by this means its paths are rendered of a considerable width. The bed of its paths exhibit, when closely examined, a curious appearance, occasioned from the erosions of its jaws, which excavate an infinity of little ramified canals. When the insect is about to assume its chrysalis state, it bores down obliquely into the solid wood, the depth sometimes of three inches; and these holes are semicylindrical, expressing exactly the form of the Grub.

At first sight one would wonder how so small and seemingly so week an animal could have strength to excavate so deep a mine, but when we examine its jaws our wonder ceases; for these are large, thick and solid sections of a cone divided longitudinally, which, in the act of mastication, apply to each other the whole of their interior plain surface, so that they grind the insect's food like a pair of millstones.

The Bombardier, or Carabbus Crepitans, of LINNÆUS, has been classed amongst the ground beetles; its antennæ are thread-shaped, and it lives mostly under ground, or in decayed wood. This insect is remarkable for its jumping motion when it is touched, we are surprised with a noise resembling the discharge of a musket in miniature, during which a blue smoke may be perceived to proceed from its extremity. The insect may at any time be made to play off its artillery by scratching its back with a needle. We are informed by ROLLANDER, who first made the experiments, that it can give twenty discharges successively. A bladder placed near its posterior extremity, is the arsenal that contains its store. This is its chief defence against its enemies, and the vapour or liquid that proceeds from it is of so pungent a nature, that if it happens to be cischarged into the eyes, it makes them smart as though brandy had been thrown into them. The principal enemy of the Bombardier is another insect of the same tribe, but three or four times its size. When pursued and fatigued, the Bombardier has recourse to this stratagem, but on the discharge of the artillery, this suddenly draws back and remains for a while confused; during which the Bombardier conceals himself in some neighbouring crevice, but if not lucky enough to find one, the other insect returns to the attack, takes hold of his head and instantly tears it off. The head, antennæ, thorax, and feet are of a brownish red colour. The eyes are black, and the abdomen and wing cases blue bordering on black, the latter are marked with broad but shallow striæ; this insect is sometimes, but rarely found in England.

[This page was blank in the original]

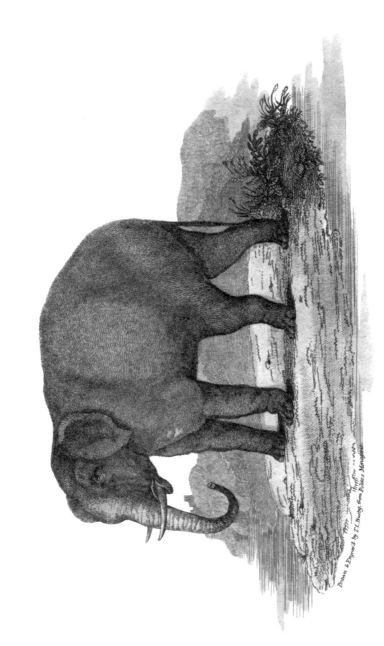

ELEPHANT.

Drawn & Engrav'd by T.L.Busby, from Police, Moorgate.

Pub. by J.Stratford, April 1 1811.

PLATE 61

ZOOLOGY.

Natural Order--MAMMALIA. Species--ELEPHAS GIGAS.

Generic Character—No fore-teeth in either jaw ; the tusks of the upper one elongated and projecting, none in the lower; the proboscis or trunk very long and prehensile; the body armed with a very thick skin, covered with a few scattered hairs.

THE Elephant may justly be considered as the largest and strongest animal at present known, and is plentifully found in a wild state in the extensive regions of Africa and Asia.

There is also found a second and different species, which is said to reside in the kingdom of Thibct, and being much smaller and of an opposite form, is to be considered as a separate animal from the above, under the title or Name of the Elephas Socotrus.

The Mammoth also may be said to have consisted of one or two species of the same kind of genus, this fact being exactly ascertained from the fossil bones and tusks which have hitherto been discovered in America, Europe, and Asia, and imagined by many writers to be the relicts of an overwhelming deluge, which at some distant period has enveloped the whole earth.

To return however to the present object, and whose history is less obscure, the Elephant may be said in general to reach the height of fourteen or sixteen feet; those of the kingdom of Siam are esteemed the tallest in respect to the breed, and vary sometimes (though very rarely) to a

I i

white or black colour. Its ears are so large, that from the shoulder of a middle sized man, they will extend to the ground; in a state of tranquillity, these are pendulous, but during the agitation of passion, they are raised and brought forward with considerable intenseness. Its legs resemble massy pillars, above five feet in height and sometimes sixteen inches in diameter.

But the most curious characteristic of the Elephant is its proboscis, which is an instrument of feeling and of motion, and which it can turn round with extreme flexibility and promptitude in every direction. With this most singular assistance it grasps every object, with both the touch and tenacity of a human finger. It is thus that it takes up herbs and roots from the ground, unties the knots of cords, opens gates and raises from the earth coins or other small objects. The trunk is internally divided into three distinct channels, reaching through the whole length, and the nostrils are situated at the end of this instrument, which is the vehicle of its food, and the weapon of its defence. In a full grown animal it is generally of the length of eight feet, and about five feet diameter in the base. In the South of Africa near the territory of the Cape, Elephants are seen occasionally in herds of several hundreds, and the settlers in the Cape Territory are frequently occupied in the employment of hunting and shooting them for the sake of the ivory, in which exercises constant practise has made them particularly skilful.

It is an occupation, however of no small adventure and peril, as the most perfect caution must be used to advance near enough to take the fatal aim unperceived, for if the Elephant observes his enemy he will rush upon him, seize him with his trunk, and tread him lifeless to the ground.

ZOOLOGY.

The general weight of the tusks of a large Elephant is about a hundred and fifty Dutch Pounds, and they are sold at least for as many guilders; so that the temptation to hunting them is not only desirable to bold and daring spirits, but to mercenary minds who can be stimulated only to exertion by the acquisition of game. The food of Elephants consists of leaves, herbs, roots, tender branches of trees, especially the plantain tree, and also of grains and fruits. It is calculated that about one hundred and fifty pounds weight of food will suffice for a single elephant for twenty four hours, but the injury which is done to plantations, when they break forwards and trample down the fences and grass is considered as for greater. They approach in concert, both for forage and mutal protection, and will lay waste in one night, all the neighbouring habitations and hopes of harvest, defying the hideous noises of the natives and the immense fires which are made to keep them off. In turning round the Elephant is slow and unweildy, although at other times his pace approaches in swiftness to the trotting of a horse; when confined within a small compass and where it is practicable he forms a considerable circle in his motion. In narrow and crooked passes the negroes avail themselves of this disadvantage and attack him with a corresponding success.

In the East Indies however, the situation of the Elephant is widely different, for being caught by stratagem, and formed gradually into the docile and obedient servant of man, his disposition becomes entirely metamorphosed by the change. In those regions of the globe, they are employed by the native princes either for the purpose of ostentation, or animals of burthen, for which latter end their back is admirably adapted. One of these creatures will execute the work of several horses, and the promptitude intelligence, and affection, they display to their keeper are singularly interesting; his voice is distinquished from all others, they will kneel down to receive him on their backs and assist him also in this operation with their trunk; with which they will frequently smooth and caress him. They are employed in drawing large caravans, and even chariots, on the most particular instances of ceremony, or festivals. The females produces but one young at a time, and that after a gestation, of two or three years. The Elephant is thirty years in attaining its maturity, and lives at the least computation one hundred and fifty years. Though they were formerly applied for the purposes of war, as was particularly the case in the time of ALEXANDER the Great, and King of Pyrrhup, yet it should seem that they were frequently more formidable to their owners than to the enemy: when wounded they exhibited a scene of extreme turbulence and confusion. At present the invention of gun-powder, appears to have precluded that advantage from their efforts in actual combat, which might in former instances have been occasionally derived from them. It is stated that in the courts of some of the Eastern sultans, they are also employed sometimes in the execution of criminals, either by trampling to death, fracturing their limbs, or impaling them in their tusks; following exactly the signals of their keeper with complete precision and correctness.

The present Elephant is delineated from a young one which is alive in the menagerie of Mr. POLITO in the Strand; he exhibits much vivacity and playfulness in his disposition, and has a curious manner of walking and standing, by placing his feet oppositely alternate, which is represented in the engraving.

[This page was blank in the original]

Drawn by G. Perry.   Engrav'd by T.L. Busby.

TROCHUS ZEBRA.

Pub. by J. Stratford March 1 1811.

PLATE 62

*Genus-*TROCHUS. *Species-*TROCHUS ZEBRA.

Generic Character—Shell pyramidically shaped, having the mouth placed underneath, leaning sideways, and of a quadrangular form, the spire inclined to the base.

IN a former number we have represented to our readers the trochus apiaria, and to which the present one is much allied. It takes its name from its outward marks, which in some degree resemble the stripes of a zebra, surrounding the spire and gradually diminishing in size to the top or summit; the general form is triangular, slanting at the base, which reveals the mouth on the underside, in a perspective view.

The turbo genus differs from the present one, in having always the mouth placed in front, and of a circular form; whereas the trochus shell exhibits the mouth concealedly underneath, with a flat ridge surround one half of the shell. The colour of the shell is of a whitish brown shaded with a dark brown, inclining to red, and the mouth of pale blue; it is probable that the form of the animal which resides within, must resemble the land snail.

It was long considered as a matter of dispute among naturalists, whether the arrangement of shells should be constituted from the animals or their habitations; no one can deny, that if we proceed upon the best principles strictly scientific, we must regard them as a department of zoology, and should on that account dispose them according to the nature and structure of the animals. But the classification formed of the characters of shells is now universally followed, and we must confess too is for many reasons preferable to any other; neither is it in the hands of the skilful conchologist, attended with so much indecision as might be generally imagined.

In the first place, amongst the vast variety of shells hitherto discovered, how small, comparatively is the number, of those, whose animal inhabitant, is described or known. It is not of species only that we speak, but of whole natural families or genera, not a single species of which has yet been discovered with the animal appertaining; so little are we acquainted with the molluscous orders, or animals inhabiting the shells. Of the shells we daily see in collections, few are fished up alive, the far greater number are found on the shores, dead or empty: neither if it were otherwise, are accurate descriptions of animals whose parts are not easily seen or anatomical investigations, which are in many cases necessary, within the capacity of every one: many of these parts and their respective functions, are not to be ascertained, except by comparative analogy, and which in itself presents an insurmountable difficulty.

Hence it becomes impossible to arrange the far greater number of testaceous productions by the animals, and the attempt must prove unsuccesful. In conchology as in any other science, the reader must necessarily acquire in the first instance a distinct knowledge of the terms employed; those which we have employed, are adopted from the authority of Linnæus, but new species being frequently discoverable in the southern latitudes of the globe, these have been found insufficient.

The distinguished members of the shells, are in general derived from the Latin, and consist of numerous names and characters.

[This page was blank in the original]

Panther.

Pub.ᵈ by J. Stratford, April 1.1811.

PLATE 63

ZOOLOGY.

Genus—FELIS PANTHERUS.
Species—PANTHER.

Generic Character—Six teeth, four central ones equal in
length, three grinders on each side, the claws
extending and retractile, the tongue armed with
prickles bent backwards; the neck being smooth,
without a mane.

THE Panther may be classed in the natural order
Felis, next to the Tiger, which animal it vies with in
the size and beauty of its skin. The gradation of form and
and colours in the Panther, Leopard, Ounce and Ocelot,
is so very gradual and imperceptible, that it requires the
nicest perception of natural history, and a close examination of the pattern of the spots, to be able accurately to
distinguish them.

The Panther is an untameable animal, and inhabits
the sultry regions of Africa and Barbary; nevertheless in
its domestic state of confinement, it is sometimes found to
exhibit a degree of greatful attatchment to its keeper, especially if captured at an early age, otherwise his natural
fierceness is too apt to break out.

The spots of the Panther are diversified beautifully
with octagonal and pentagonal marks, darkest towards the
back, exhibiting a variety of effects highly delightful
to the eye, the ground being of a pleasing buff colour, with
the exception of the ears and lower part of the body which
are entirely white.

The Tiger on the contrary is distinguished with dark
stripes continued round the body

ZOOLOGY.

The Leopard has spots of brown surrounded with circular rings of the same colour and gradually diminishing as they descend forwards the feet, and the whole has a darker effect than in the Panther tribe.

In the boundless tracts and impenetrable forests bordering on the Senegal and Congou Rivers, the Panther lurking under the thickets generally seizes his prey by surprize; or by creeping on its belly till it comes within reach. When pressed with hunger it attacks every living creature without distinction, but happily prefers the flesh of brutes to that of mankind; it will even climb up trees in pursuit of Monkies and lesser animals, so that nothing is secure from its attacks.

The size of it is something more than a large Mastiff Dog, but its legs are not quite so long. Its voice is strong and hoarse, and it growls continually.

The Ancients were well acquainted with these creatures, for we find from the Roman History that they drew prodigious numbers from the Desarts of Africa for their public shows; sufficient one might have supposed to have entirely exhausted them. Scaurus exhibited an hundred and fifty of them atone time; Pompey four hundred and ten; Augustus four hundred and twenty. They probably thinned the coasts of Mauritania of these animals, but they still seem to swarm in the southern parts of Guinea.

The present Panther is delineated from an excellent living specimen in the Menagerie of Mr. Polito,

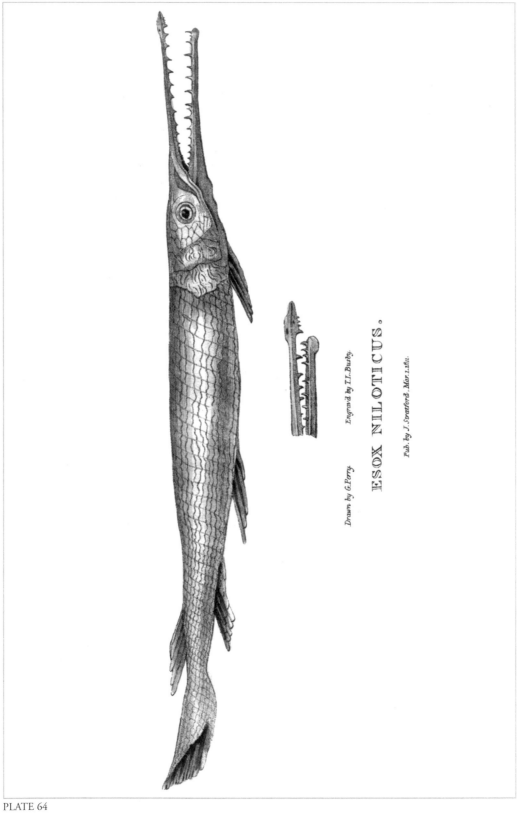

Drawn by G.Perry.   Engrav'd by T.L.Busby.

ESOX NILOTICUS.

Pub. by J. Stratford. Mar.1.1811.

PLATE 64

Genus—ESOX. Species—ESOX NILOTICUS.

Generic Character—The top of the head flattened and ending
in a very long upper jaw, thickened at the end,
having two small spiracles or breathing holes, one
placed above, one below the eye-socket, of an
oval form ; the body and part of the tail curiously
plated with shining scales ; the tail fibrous and
square at the ends, the dorsal fin placed opposite
the anal fin and near the tail, the upper mandible
of the jaw projecting like a spoon, and armed with
teeth.

IN the form and character of the present curious fish,
which was caught in the river Nile, and brought to England
by Lieutenant FINDLAY; we offer many curious characters
to the eye of the Naturalist, and it is difficult to pronounce
whether the head, beak, body, or tail, present the most
striking singularity ; there is certainly some distant analogy
to the crocodile, or the boney scaled pike of America ;
from which however as well as from the common English
pike, and the various tribes of sword fish, it differs considerably. It may perhaps be regarded as a sort of intergenus
or gradation amongst the various tribes of predacious fish,
and is calculated rather to excite attention by its form than
its colour, being naturally of a pale and pearly appearance.
The eye is placed very low in the head, and the mouth
is rather narrow ; these two circumstances seem to indicate
that the prey which it devours must be chiefly of the
smaller kind of fishes, for it seems not to have the power for
piercing larger objects, owing to a blunted form of the
upper beak, and which it would be difficult afterwards to
extract.

K k

When his Majesty's ship the Leopard, after her arrival from the Coast of Guinea, was ordered in 1725, to be cleaned and refitted for the channel service, in stripping off her sheathing, the shipwrights found in her bottom, pointing in a direction from the stern towards the head, part of the sword or snout of one of these fishes, and the substance where it was broken off, appeared like a coarse kind of ivory; it had penetrated through the sheathing, which with seven inches of solid oak timber, was pierced by the solid fragment: the workmen on the spot declared it impossible, with a hammer of a quarter of an hundred weight, to drive an iron pin of the same form and size, without nine or ten blows, though this was effected only by one. Another instance of the same kind has since occurred, in which the sword fish was embedded, or driven through for the whole length, and the fish was supposed to have been killed by the violence of the effort; part of the vessel and the sword are now deposited in the British Museum.

The sword fish and the whale, are said never to meet without coming to battle, and the former has always the repute of being the aggressor. Sometimes two of them join against one whale, in which case the combat is by no means equal; the whale uses his tail only for his defence, he dives down into the water head foremost, and makes such a blow with this instrument, that if it takes effect, it finishes the sword fish at one stroke; but the other who is in general sufficiently nimble to avoid it, immediately falls upon the whale and buries his weapon in his side. the whale perceiving the sword in his side (but usually it only pierces the blubber) darts again to the bottom, and thus the conflict is renewed with various success, until one of them is discomfitted by the superior strength of that of his enemy.

REMARKS

ON THE INFUSORIA, INSECTS, AND POLYPI,

DISCERNABLE ONLY BY THE MICROSCOPE.

It may be truly observed in respect to the great discoveries of natural history by means of the microscope, that so numerous are the minute, but wonderful tribes of animals, which it presents to our view, that he is only acquainted with one half of the creation of the great artificer, who has not joined some knowledge of the animalcules to his other views of nature.

The works of BONNET, LEUWENHOEK, BAKER, and other labourious philosophers, have convinced us that millions of animals smaller than the eye can see, or even the mind conceive, exist in every object of the creation, which is either fluid or sold. The Multiplication of these is produced in a manner which is still hardly understood, but in certain kinds as the **Polypus**, **Vibrio**, and others of the same class, should the parent be divided into several small parts with a knife or other means, each part being placed again in the water (which is their natural element) becomes instantly a living creature. Others seem naturally to divide themselves into a regular number of sections, and each part separately forms a new animal similar to the parent.

The polypus is a small gelatinous animal, living in fresh water, and is amongst the most wonderful productions of nature. The particulars of its life, its mode of propagation and power of reproduction after being cut to pieces, are truly astonishing to a reflecting mind. The green polypus is formed like a small bell, suspended by a membrance at the base to some other body or leafy substance; it has eight or ten thread-like arms which surround the mouth in equal divisions, and which it uses in seizing its food. It is capable of contracting its body in a very sudden manner when disturbed, so as to appear only like a round spot; and when the danger is over, it will expand itself again to its former shape. It may be cut in every direction that the fancy can suggest, and even into minute divisions, and not only the parent stock will remain uninjured, but every section will become a perfect animal. Even when turned inside out it suffers no material injury, for in this state it will soon begin to take its food, and perform all its other natural functions.

Mr. Trembley of Genera, discovered in his experiments, that different portions of one polypus could be engrafted one on another, two transverse sections brought into contact will soon unite and form one animal, although each section belong to a different species. The head of one species may be grafted on the body of another. When one polypus is introduced into the others body, the two heads untie, and form one individual, thus by repeatedly splitting the head and part of the body, a complication of hydras may be formed more wonderful, than ever struck the imagination of the most romantic fabulist.

These little creatures continue active during the greatest part of the year, and it is only when the cold is most intense, that they feel the general torpor of nature; all their faculties are at that time for two or three months suspended; but if they abstain in one, they have ample amends in their voracity in another, and like those animals that become torpid in winter, the meal of one day suffices for seven months.

The eel vibrio is an excessively minute insect, resembling an eel or worm in its form, and is found in sour paste diluted with water, or in vinegar, they bring forth their young ones alive, and produce a numerous progeny. If one of them is cut across through the middle, several young ones coild up and enclosed each in a membrance will be seen to issue from the wound, which readily accounts for their sudden prodigious increase.

The Proteus Vibrio is a species that has its name from its very singular power of assuming different shapes, so as sometimes to be with difficulty taken for the same animal. When water, in which any animal substance has stood undisturbed for some days, a slimy substance will be found to contain, among several other Animalcules the Proteus. It is pellucid and gelatinous swimming about, most comonly with a long neck and bulbous body, with great vivacity; sometimes it makes a stop for a minute or two, and stretches itself out apparently in search of prey; when alarmed and immediately draws in its neck, becomes more opake and moves very sluggishly. It will then perhaps instead of its former long neck, push out a kind of wheel machinery, the motions of which draw a current of water and along with this probably its Prey. With drawing this it will sometimes remain almost motionless for some seconds, as if weary; or instead, adopt in succession a multitude of different shapes. The eyes of this animal have not hitherto been discovered; it swims however with great rapidity amongst the multitudes of animalcules that inhabit the same Water without ever striking against them.

The Volvox Bulla is described as a species of minute creatures, the most simple of any that have yet come to our knowledge. They consist only of oval bodies; similar in appearance to soap bubbles arranged in parties of three, five, six, and nine, among them are also some solitary ones. These collections of Globules, being put into a glass filled with Sea-Water, described a rapid circle round the glass by a common movement, to which each individual contributed by the simple compression of the sides of its body, probably the effect of the reaction of the Air with which they were filled.

The Animals denominated Zoophytes by Linnæus, and most other naturalists, seem to hold a middle station between animals and vegetables. Most of them deprived altogether of loco-motion are fixed by stems that take root in crevices of Rocks, among Sand or in such other situations as nature has destined for their abode; these by degrees send off branches till at length some of them, attain the size and extent of large shrubs.

The Zoophytes are usually considered under two divisions. The stony branches of the first division which has the general appellation of Coral are hollow and full of cells, the habitations of animals resembling Polypes, star fish, &c. according to their respective Genera.

They consist of the Tubipores, Madrepores, Millepores and Cellepores and are nearly all confined to the Ocean.

The animals appear at the ends of the branches, where they have somewhat the resemblance of animated blossoms, endowed with considerable spontaneous motion.

The stems of some of the Millepores are almost solid, the cells being so extremely small as to be scarcely visible without high magnifying powers. Amongst these is the Millepora Polymorpha or officinal Coral of the shops.   The next division of the present order consists of such animals as have softer stems and are in general not merely inhabitants of a stem and branches, but are themselves in the form of a plant.

Those best known are the Corallines, the sponges and the Polypes.   The Corallines are composed of capillary tubes whose extremites pass through a calcareous crust and open pores on the surface.   They are entirely sub-marine, and from their branches being finely divided and jointed, resembling some species of Lichens, they have till late years been arranged by Botanists with the Cryptogamous Plants.   In appearance they certainly approach very nearly to some vegetables, but their calcareous covering alone is sufficient demonstration of their being allied, (though at whatever distance it may be) to the animal Creation

The Sponges consist of an entirely ramified mass of elastic capillary tubes, supposed by many to be the production of a species of worms, which are often found, straying about their cavities.   This idea is however now nearly exploded, Others have imagined them to be merely vegetables.   But that they are possessed of a living principle, seems evident from the circumstance of their alternately contracting and dilating their pores, and shrinking in some degree from the touch, whenever examined in their native waters.   From their structure they are capable of absorbing nutriment from the fluid in which they are by nature immersed.   They are the most torpid of all the Zoophytes; the species differ greatly from each other, both in shape and structure; some are composed of reticulated fibres or masses of small spines, some are as the common sponge of no regular shape others cup-shaped or tubular &c.

The Officinal Sponge which is sold in the shops is elastic and very full of holes, it grows into irregular lobes of a woolly consistence, and generally adheres by a very broad base to the rocks. It is chiefly found about the Islands of the Mediterranean Sea, where it forms a considerable article of Commerce. A variety of small marine animals pierce and gnaw into its irregular winding cavities. These appear on the outside, by large holes raised higher than the rest. When it is cut perpendicularly, the interior parts are seen to consist of small tubes, which divide into branches as they appear on the surface. These tubes, which are composed of reticulated fibres extend themselves every way, by this means encreasing and ending at the outside, in an infinite number of small holes, elastic and compressible, and which are the proper mouths of the animal. Each of these holes is surrounded by a few erect pointed fibres that appear as if woven in the form of little spines. These tubes with their ramifications, in the living state of the sponges are clothed with a gelatinous substance, properly called the flesh of the animal. When the sponge is first taken it has a strong fishy smell and the fishermen take great care to wash it perfectly clean in order to prevent its subsequent putrefaction.

To enumerate all the different forms of the Coral Genus, by their own peculiar names, would far exceed our present limits; we therefere refer our Readers to Mr. ELLIS's Work upon the Corals and Corralloids, the first writer that ever illustrated this difficult and elaborate subject. It contains however such a number of beautiful objects for the Painter and the Artist, as will amply repay the trouble of imitation and unfold even in the most cultivated taste, new ideas of the beauty of the creation.

[This page was blank in the original]

Drawn & Engrav'd by T.L.Busby

LEOPARD.

Pub.d by J. Stratford. May 1.st 181.

PLATE 65

[448]

Genus—FELIS. Species—FELIS LEOPARDIS.

Generic Character—Six teeth, four central ones, equal in length, three grinders on each side, the claws retractile and extending, armed with prickles bent backwards; the neck smooth without a mane,

THE Leopard is very nearly approximated to the Panther, which we have described in a former part of the ARCANA, so as to make it difficult to distinguish the two creatures. The former seems to have taken its name from a Latin derivation Leo-pardus, signifying a spotted lion, as having a considerable resemblance to the lioness in the general form of its head and limbs. This animal is smaller than the Panther, and each of the spots on the body are composed of five black points forming a sort of circle with a black circular spot of the same size in the center, varying however in shape towards the feet and under-side of the body. The colour of the ground is in general of a lively yellow, the underside of the body verges suddenly to a white colour. His manners and habits may be perfectly subdued by his keeper, and exhibit a much more favorable disposition than the Tiger or the Panther. All these species, of which the Wild Cat of England and France may perhaps be placed at the lower end of the scale, seem to degenerate in fierceness, as they decrease in size, as if conscious of the superior powers of man. The rivers of Senegal and Niger are the most remarkable parts of Africa for the growth of the Leopard. The Negroes and Arabs take them in pitfalls slightly covered with twigs at the top and baited with flesh, and at other times hunt them with spears, when they crouch low with cowardly fear from the sight of their numerous enemies, and are readily taken by a rope or net. The skins are reckoned very beautiful and are brought to Europe to cover saddles, &c. where they are highly esteemed.

B

ZOOLOGY.

In India there is a species of Leopard that is of the size of a Greyhound, with a small head and short ears. Its face, chin, and throat are of a pale brown colour inclining to yellow, the body is of a light tawney brown; this animal is marked very richly with small round black spots scattered over the back, sides, head, and legs; the hair on the top of the neck is longer than the rest; the belly is white, the tail very long, marked on the upper side with large black spots. This is undoubtedly to be considered as a distinct species from African and Syrian Leopards, and which has been for many ages made use of in that country for hunting the Antelope, as well as other beasts of chase. It is carried in a small kind of waggon, chained and hood-winked till it approaches the herd, when it is released and suffered freely to prosecute its natural pursuit of game. It begins by creeping along in a lurking attitude close to the ground, and concealing itself until it gets an advantageous situation, it then darts on its prey with great agility, frequently making five or six amazing bounds. If it should not succeed in its first effort, it gives up the point for that time and readily returns in a submissive manner, to its master. Thus the savage powers of the wilder animals are bent to the useful or amusing occupations of man; and even the smaller voracious creatures, as the Hound, Cat, Ferret and Weasel are found to materially add to his powers by their useful services and assistance; whilst the gregarious tribes of beasts being generally less capable of labour, and affording him a more agreeable food, willingly yield up to him their existence, as a return for the increased pleasures which they most generally receive from him, and which ought certainly to soften his disposition to gentle and humane sentiments; and lead him to take from them their life with as little pain as possible.

[This page was blank in the original]

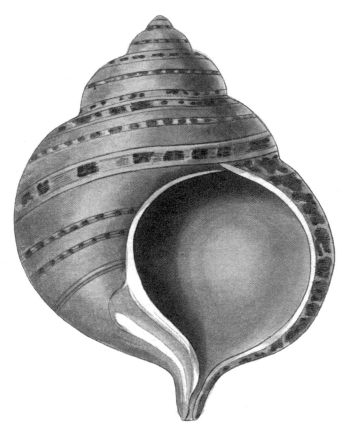

Drawn by G. Perry.   Engrav'd by T.L. Busby.

BUCCINUM DISTENTUM

Pub. by J. Stratford, May. 1. 1810.

PLATE 66

[452]

CONCHOLOGY.

*Genus—*BUCCINUM.
*Species—*BUCCINUM DISTENTUM

*Generic Character—*The shell, ovate or elongate; the cheek
standing straight forward, the opening oblong;
the beak invested at the back with a thick twisted
collar swelling outwards, no canal; the body
covered with tubercles and continuous bands.

IN the present number we have offered to the view
of our readers a shell from the southern regions of the
globe, and which we imagine to be hitherto undescribed by
any recent author of note, in the article of Conchology.
It differs from others chiefly in the shortness of the rostrum
or beak, and more especially from that which we have
described formerly in No. XV. of the ARCANA, in having
none of these circular projecting ridges which are placed
so numerously upon the spire, body and beak of that
Buccinum. The spiral folds are three in number, regularly
diminishing from the summit of the body, which forms
with the opening of the right side, a very great tendency
to a circular or globular shape; the edge of the mouth is
shape and thin, the beak sharped like a spout, the inside
being of a blue colour; the predominating colour of the
body is of a pale pink enlivened by bands of yellow and
brown of various breadths and tints, diminishing towards
the top with a beautiful glazing or varnish, which serves
to adorn it in a way that baffles all the imitation of painting.

In the minute or microscopic tribes of shells of the
Buccinum kind, we find the same reasons for wonder and
astonishment of the mind, as they equal in elegance and
variety the larger species, and are supposed to exist in
a much larger quantity, and found to be in general of
similar texture and construction. They exhibit through

[453]

a magnifying-glass a vivacity of colour and pearley appearance hardly to be equalled by any other object in nature.

We have imparted to our readers in a former number, how short and confined our knowledge is, and will always be, respecting the internal structure of the shell-animals. There is certainly a material difference in the construction of the univalves and bivalves, the former are furnished with an operculum, door, or lid, which it can, by a muscular action, draw in close, so as to defend itself completely from its enemies, it being thick and as strong as ivory or bone, and fitting closely in the cavity like the lid of a box. The bivalves, on the other hand, have no instrument of this sort, but as they live in a double shell, they have a muscular motion within, by which they can closely bring the parts together, and secure themselves from all danger. This is sometimes performed by a hinge as in the Donax, and in other cases by a muscular strong sinew, which yields only to the knife. They have a stomach which is single and flat, following the form of the shell in a certain degree; when hungry they open cautious by the two sides, and like the Polypus, live chiefly upon small insects and plants. The univalves have expansile horns like Snails, which differ in number and form, being capable of such a great degree of contraction, as to become invisible or not, just as the animal pleases: these Tentacula are supposed to have been constructed for the purposes of drawing in their food by a circular motion, and also have the eye placed at the end which they can open and shut at pleasure. The situation of the mouth is not at present known, and many of the species seem to live entirely without it, and perhaps find a substitute in a state of external absorption. From the experiments of Spallanzani, we may conclude that they absorb much oxygen externally.

[This page was blank in the original]

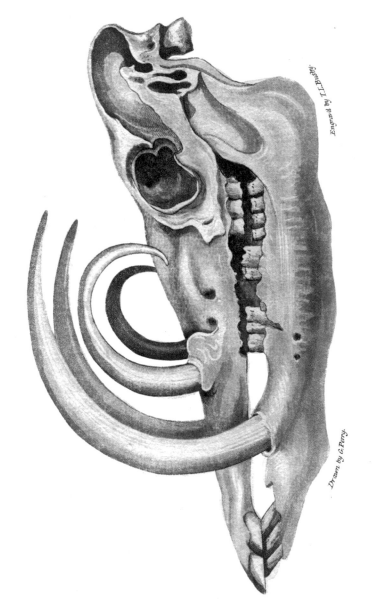

Drawn by G. Perry.

Engrav'd by T.L.Busby.

B A B Y R O U S S A.

Pub. by J. Stratford, May 1.1811.

PLATE 67

[456]

Genus—BABYROUSA. *Species*—QUADRICORNUA.

Generic Character—Four tusks projecting, two in the upper
      jaw, small and circularly reverted upon the cheek,
      two others, large and enclosing the first, termi-
      nating each of them in a blunted point. The
      outer tusks are round, and the greatest part of
      these not gradually, but suddenly terminating;
      the inner ones more slowly diminishing, and the
      external parts in a small degree flattened at the
      extremities.

THE present curious skull, of a most singular form,
belongs to an animal very seldom seen alive in England; it
resides near the countries of Amboyna, and is so nearly
similar in the character of its head and tusks to the fossil
bones found in North America and Siberia. These latter
are so different from any animal at present existing,
that we shall take an early opportunity of making a comparison from such specimens of fossil remains that present
any uncommon and striking analogy with those which are
now existing. It has been supposed, that two species of
the Mammoth existed in former ages of the world, and this
opinion seems to be confirmed by the following circumstances. The pointed grinders which are found in a fossil
state are cuspidated on the upper edge, and enamelled;
others are found circumossated with an undulated surface,
flat and of an angular plan. The Babyrousa has in its front
teeth a particular similitude to the Jerboa or Kanguroo, in
the slanting direction of the fore teeth; and in its teeth we
imagine it will be found, as well as in its tusks, similar to one
of the species of the Mammoth.

c

ZOOLOGY.

The Tapir and Babyrousa have a very considerable analogy of form and character, the former is found in South America, near the isthmus of Darien, and the river of the Amazons. It is said to have five hoofs on each foot, and ten cutting and ten front teeth in each jaw. The Babyrousa has ten teeth in the front of his mouth placed in a slanting manner, the number of the grinders is not exactly known, as authors vary in their account. The grinders stand far behind in the back of the mouth, and the canine teeth are very small, but are unfortunately lost in the present specimen. It is impossible to say for what use the tusks are formed in such a strange manner, as it seems as if only two of them could be used, and these are the outer ones; the under ones being placed against each cheek and turning out a little from the head, could hardly be employed for weapons of defence or attack. There seems to be a considerable mistake in the position and form of the horns, in the wood engraving of Mr. BEWICK, who also classes it with his predecessors amongst the Hog tribe; the only similarity seems to be on account of a likeness in the form of the snout. If indeed the species had been seen alive, which we do not find to have been the case, there must be two species, since those tusks do not at all accord with the present specimen. We have already noticed the singular form of the hoofs of the American Tapir, which is a light and elegant animal, and will be found perhaps to be more analogous to the Babyrousa than any other. It would not be very unreasonable perhaps to take the system of KLEIN, placing the distinctions in the form of the feet, the Tapir then would be entirely removed, and several contradictions as to the Sloth tribe and others.

[This page was blank in the original]

GUANACO.

Pub. by J. Stratford, May.1.1811.

PLATE 68

ZOOLOGY.

Genus—GUANACO.　*Species*—GUANACO, PATAGONIA.

Generic Character—Fore legs bending forwards, feet cloven
and pointed, the tops of the feet deeply divided,
the back rounded and covered with long brown
wool; the breast and shoulders with very white
hair, the ears long, white and pointed forwards, no
horns, the face resembling the Antelope, part of the
hair of the neck, under the ears, soft like silk.

AMONGST the various animals which are peculiar to
South America and are found no where else, may be reckoned the Guanaco, whose existence has been almost disputed
by European naturalists, by being confused with the descriptions of the Lama and Vicuna; and the drawings made
by the Spanish naturalists being very indifferent and contradictory. Even Dr. SHAW himself, who quotes GESNER, and
gives the name adopted by SCHREBER, of Allo, Camelus,
Cervus, Huanacus, &c. seems to be in doubt whether it is
not merely a variety of the Vicuna or Lama. GESNER described a specimen as being alive in 1558, and informs us
that the neck was all over white as a swan, and that the
body was all of a dark purple brown, cheeks and legs' ex-
cepted. This seems to be the first good description, as to
the colour of the creature. The true Lama has never been
seen alive in England, and is, truly speaking, a Peruvian
animal, distinct from the Vicuna, which is now common
enough in England. The menagerie of animals, as at present established at Paris, is in possession of two real Lamas,
which were brought over alive as a present to Madame Buo-
naparte, so that the three species may now be said to be
clearly ascertained. The Guanaco varies in size from four

to five feet, the other proportions will be perhaps best discerned from the drawing, in an animal which is thickly covered with so much flowing hair; the ears are taper and of a form partly resembling the ass, the tail is short and bushy, sometimes quite concealed; but the greatest singularity is the projecting position of the fore legs, in which the knees are depressed inwards. There is a natural separation of the hair of the body from that of the thigh; the face with the ears are rather grey, and help to set off the silky, downy white of the neck and mane, which are curiously blended into one mass. Of the habits and disposition of this creature we must speak much more favourably than of the Vicuna, as it never spits in the face of strangers, as the latter is very apt to do. To its keeper it displays a docility and gentleness which generally accompanies the hornless animals. The eyes are perfectly black and there are brown spots and two black ones upon the front shins, that are very remarkable, as well as several small ones scattered upon the face. The characters, however, of the Guanaco are so various, that we hardly know to what animal he is different, having the hair of the Syrian Goat, the feet of the Vicuna, the ears of the Ass, and back and tail of the Sheep. The softness of the under wool of the neck, seems to render it very fit for the manufacture of hats, whilst the coarser kind might serve for rugs, coverlids, muffs, and other convenient purposes. The breed is said to be most numerous in the mountains of Patagonia.

This curious animal is drawn from a living specimen now (April 1811) exhibiting in Piccadilly.

General Remarks on the King-fisher Tribe.

THESE birds frequent the banks of rivers, living principally on fish, which they catch with great dexterity, they swallow their prey whole, but afterwards throw up the indigestible part, from a peculiar conformation of the stomach. Their wings are short, and yet they can fly with very great swiftness.

The common King-fisher, the Alcedohispida of Linnæus, has been denominated the most beautiful of the native birds; although we conceive the Jay, and the Silk Tail, or Waxen Chatterer, to be quite equal; a large bill and very minute legs, seem to be remote from our general ideas of symmetry, the legs are red, the bill is black, but yellow at the base, the rest of the colours are a rich azure or sea green colour intermixed with blue, the tail of a very dark blue.

This bird is found in all parts of Europe and the north of Asia. It preys on small fish which it finds in the broadest of the trout streams, where it may be seen swiftly passing each way, like a blue ray of light. It frequently sits on the bough of a low bush, overhanging the current; there it remains immoveable, until it sees a small fish springing to the top, he then dives in the water and generally brings up his prey, which he carries in his bill to land, and after tearing its flesh, swallows it whole.

When the King-fisher cannot obtain a projecting bough, he sits attentively on some projecting bank near the brink; the moment it perceives the prey, it mounts into the air for ten or twelve feet, and drops perpendicularly from that height. It is often observed to stop short in its rapid course and remain stationary, hovering like the Hawk tribe over the same spot for several seconds. This is its mode in Winter, when the muddy swell of the stream, or the thickness of the ice, constrains it to leave the broader rivers, and traverse the unfrozen brooks. At each pause it continues, as it were, suspended at the height of fifteen or twenty feet, and upon wishing to change its place, it skims along about a foot from the surface, and he halts and rises again.

This continual exercise, so often repeated, seems to shew that he dives for many small objects, fishes and insects, and often in vain, for he passes for many miles in this way. The King-fisher, we are informed by Gmelin, are seen all over Siberia; their feathers are used by the Tartars and the Ostiacs for many pious uses. The Tartars pluck them, cast them into the water, and carefully preserve such as float; such is their foolish superstition, that they suppose if a woman touches these, even with her cloaths, she must fall in love immediately. The Ostiacs take the skin, the bill, and the claws of this bird, and shut them up in a purse; and as long as they preserve this sort of amulet, they believe that they have no ill to fear. The person who wished to teach Monsieur Gmelin this means of living happy, could not forbear shedding tears, he told him that the loss of a King-fisher's skin had caused him to lose also his wife and his goods. Monsieur Gmelin observed such a bird could not be very rare, since a countryman of his had given it to him with all its skin-feathers; he seemed very much surprised, and informed him that if he had the good luck to meet with one, he would certainly part with it to no one.

Mr. Daubenton kept these birds for several months, by means of small fish placed in basons of water on which they fed; for, on experiment, they refused all other kinds of nourishment.

The King-fisher lays its eggs to the number of seven or more, in a cavity on the banks of the river it generally frequents. Dr. HEYSHAM had a female bird, brought alive to him at Carlisle, by a boy, who said that he had taken it in the preceding night, when sitting on its eggs. His information on the subject was, that having often observed these birds frequent a bank on the river Peteril, he had watched them carefully, and at last saw them go into a small opening in the bank; the hole was too narrow to admit his hand, but as it was made in soft mould, he easily enlarged it. It was upwards of half a yard long; at the end of it were the eggs, which were six in number, were placed with care upon a substance here-after to be described, having very little appearance of a nest. The eggs were considerably larger than those of the yellow-hammer, and of a transparent white colour.

It appears, from a still later account than the above, that the direction of the hole was always upwards, and have there, inclosed at the farthest end, which is the widest, a kind of bedding, formed of the bones of small fish and some other substances, evidently the castings from the stomach of the parent animal, and evidently brought here for the construction of a very singular kind of nest. It is generally kneaded like a hollow cake, about half an inch in thickness, and mixed with a portion of earth and other substance, as the dead leaves of grass, &c. to make it bind. There is every reason to believe, that both the male and female come to these spots, for the purpose of ejecting their food, for some time before the female begins to lay her eggs; and that they warm it and dry it by the heat of their bodies, and raise it to a sufficient height above the water's edge, as a security against the floods.

They have been known to remain frequently for two or three hours in order to accomplish this purpose, and long before the period of laying. When the young are nearly full feathered, they begin to shew a strong appetite, and exhibit such a strong desire of food by their chirping noise, which sometimes leads to the discovery of their secret abode.—— What strange instinct leads the parents to succour them when young and helpless, and afterwards to urge them to efforts of flight and diving in a region as yet unknown? What kind regard is it that impels the tender mother to urge its half-fledged progeny to the steep precipice, and at last pushes its trembling wings into the wide expanse. Instinct to be unkind to that which they can no longer lead, and perhaps the power of memory suggests to the older one its former experience: imitation in the young birds, must also be a strong incitement to seek for an enlargement of action. It has been well observed by some Philosopher, that we can see, in these cases, nothing less than the Divine Power itself, operating immediately upon the lower class of all the animated creation.

[This page was blank in the original]

G.P. del.

T.L.B sc.

SPHYNX CASTANEUS.

Pub.ᵈ by J. Stratford, June 1.1811.

PLATE 69

Genus—SPHYNX.

Species—SPHYNX CASTANEUS.

Generic Character—The wings long and narrow, the antennæ thickest in the middle, sharply pointed and bent down at the ends, the body divided by joints, and partly invested with a hairy covering, the fore legs projecting, and armed with a three-fold claw, the lesser antennæ curvated and pointed outwardly; upper wings smooth at the edges, and round at the ends, the lower parts of the wings undulated.

AMONGST the various Insects which have been discovered in New Holland, a magnificence of size, and rich splendour of colours, in the moth, butterfly and Sphynx tribe have attracted the attention of all the voyagers who have visited these curious regions of the globe.

The Sphynx genus differs from the Phalæna and Papilio chiefly in the form of the Antennæ, and the great length of the body extending far beyond the wings, the head also is more elevated; these circumstances serve to ballance it in its flight through the air. Like the butterfly of England and silk worm of China it undergoes four stages, or changes of existence, from the egg to the caterpillar, afterwards to the chrysalis or enclosed state, and lastly its flying existence endowed with wings, and presenting to the inquisitive mind of man an emblem of the immortality of the soul; and their numbers are again renewed in the multiplication of their infant brood, deposited as eggs, upon various vegetable productions, and followed shortly by the death of the parent fly.

L l

ENTOMOLOGY.

The Sphynx Castanea is a native of Port Jackson and Van Diemen's Land, and exceed most of its congeners in size and beauty; the upper are of a reddish brown, edged with black and white shades, the body is covered with a soft green down and ending in a jointed tail of a yellow colour; the under wings are of a bright chesnut tint on the upper part, shaded below with brown and black, from which circumstance we have given it the name Castaneus. The whole fly (except the antennæ and tail) consists of feathers inserted in a membrane, and resembling when examined carefully with a microscope, the wings of a bird, thus exhibiting to our view a degree of workmanship and contrivance which baffles our inmost thoughts, convincing us of the wisdom of the divine hand that first formed them. No art can rival, no painter equal the arrangement of the colours, so that we may exclaim with our excellent poet, THOMPSON:

> —————"Who can paint like Nature!
> Can imagination boast amidst its gay creation,
> Hues like hers; or can it mix them with that
> Matchless skill, and lose them in each other,
> As appears in every Bud that blows."

or with the divine MILTON;

> "These are thy glorious works, Parent of Good!
> Almighty thine this universal frame,
> Thus wond'rous fair, thyself how wond'rous them—
> Unspeakable who sitts't above, unseen in these
> Thy lowest; yet even these thy lowest works declare
> Thy goodness beyond thought and power divine."

[This page was blank in the original]

C. Hirst del.

Pub.d by J. Stratford, June, 1.1811.

PLATE 70

[472]

Genus—LANIUS.
Species—LANIUS AURANTIUS.

Character—The Tail convex and consisting of feathers. The Beek hooked and the upper bill doubly incurvated; the feet resembling those of the Hawk bnt not so long or slender in its proportions of the different parts of the body and bill.

THE Genus Lanius, is commonly called the Shrike or Butcher Bird, is very naturally and properly placed by Linnæus between those which are partly rapacious as the Crow Tribe, and such are much smaller as the Passerine or Swallow, forming a connecting and divisionary link between these two families. The several species of the Shrike are found in every quarter of the globe: except within the Arctic circle, and present an astonishing variety of form, size and colour; although a bird of a diminutive appearance it will readily seize upon its prey with the greatest voracity and boldness. The Muscles that move the bill of the Shrike are very thick and strong, an apparatus peculiarly necessary to a species whose mode of killing its prey is so singular, and whose manner of devouring it is so extraordinary. He generally seizes the smaller birds by the throat and strangles them, and for this reason the Germans call him by a name, signifying the suffocating Angel. When his prey is dead, he fixes it on some thorn ; and thus spitted, tears it to pieces with his bill. Even when confined in a cage he will often treat his food in the same manner, by sticking it against the wires before he devours it.

In spring and summer, he imitates the voices of other birds, and by this means decoys them into his reach regarding them as his natural prey. Except in this instance, his natural notes are much the same, through all the months of the year.

In Mr. Bell's Travels from Moscow to Pekin, we are informed that the Russian bird catchers often take the Shrikes home for the purpose of taming them. He had one of them given to him, which he taught to perch on a sharpened Stick, fixed in the wall of his apartment. Whenever a small bird was let loose in the Room, the Shrike would immediately fly from his perch, and seize it by the throat in such a manner as almost in a moment to suffocate it. He would then carry it to his perch and spit it on the end which was sharpened for the purpose drawing it on carefully and forcibly with his bill and claws. If several birds were given to him, he would use them all one after the other, in the same manner. These were so fixed that they hung by the neck till they had leisure to devour them.

The present Shrike of which we have here given our readers a representation, is kindly communicated from the Museum of Mr. PRIESTNALL, and something less than the common Thrush. It is a native of Buenos Ayres, and we believe it to be hitherto wholly undescribed. The back of the head, neck and tail is of a black colour, the body and vent feathers of an orange; the wings touched underneath with the last mentioned tint. The legs are of a pale brown, the back adorned all over with very ovius white spots, regularly placed. The bill very large in proportion to the size of the bird, and of a blue colour, the wings small, and the whole combining a very pleasing assemblage and oonstituting one of the most striking species of this most predacious but elegant family of Birds.

[This page was blank in the original]

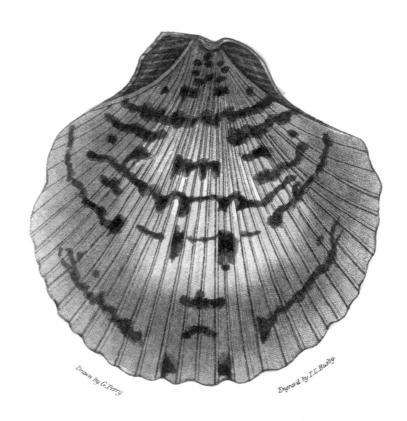

Drawn by G.Perry

Engrav'd by T.L.Busby.

PECTEN SANGUINEUM.

Pub.d by J.Stratford, June.1.1811.

PLATE 71

Division—BIVALVE. *Genus*—PECTEN.
Species—PECTEN SANGUINEUM.

Character.—Shell round and flat, armed at the hinge with either one or two angular wings or flaps, the ribs diverging in a radiated form serve to distinguish it from the oyster and all other of the bivalve shells.

THE division of multivalves has been judiciously excluded from the other parts of the system of Conchology, as they did not easily assimilate in the Analysis or habitudes of existence. The two great orders or divisions may be said of two parts, first the univalves or spiral shell, incapable of widening its mouth and having a circular flap by which it can close up its shell. The bivalve, of which we have hitherto given no exact explanation or drawing, is now elucidated by the Pectum Sanguineum which consists of two shells of equal size joined by a hinge. The body of the fish which resides in this shell is of a flat form and capable of attaching itself by a tendinous muscle to rocks or other substances, and is completely preserved and shut up from its enemies when the shell is closed. The spiral or univalve retires like the snail, by contraction, and draws close; the round operculum which resembles the lid of a box, and is of a hard boney nature thickened in the middle. Thus each is provided with a providential defence, and if their shells should be unfortunately broken, they have a viscous fluid by which they can repair the damage, that secreted and applied from the tongue and settles into a hard matter resembling the original shell.

CONCHOLOGY.

The **Pecten Sanguineum** is rather rare as a shell, and is generally found in the Red Sea and eastern coast of Africa. Its form is resembling an ancient comb, from which circumstance it takes its name, it has pointed wings of different sizes and form, placed on each side of the hinge, from these rays all regularly diverge and enlarge themselves in right lines like the beams of the sun. These animals do not live together like oysters, but are cast up in a solitary manner upon sandy shores or in shallow bays. Their flesh, especially of the foreign ones, is deemed a very nourishing and delicious food when boiled or roasted, this is provided by sailors when an opportunity offers as a powerful remedy for the scurvy or flux. The size of the Pecten is seldom exceeding eight or nine inches, but the smallest are astonishing for the beautiful variety of spots which abound in them, placed in various groupes amongst the ribs which environ the shell, the ribs being very strongly indented on the inner side, has presented to the painter and carver, an object of pleasing and picturesque character. Thus from the mathematical forms of natural objects, though some are loathsome in their forms and proportion, yet these generally avoid the haunts of men, and seem admirably by their contrasts to be fitted to make us more contented, by considering that many of their uses may be unknown to the human eye, and if ever discovered prove a fresh source of thankfulness in man.

[This page was blank in the original]

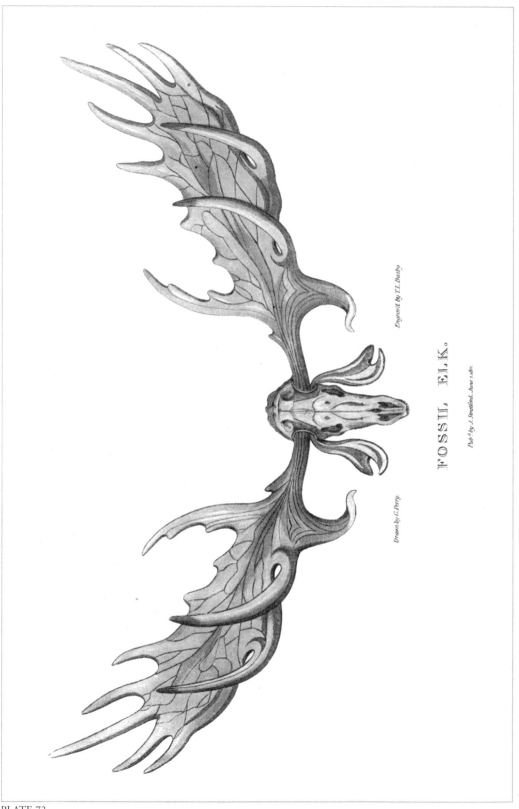

Drawn by G. Perry.

FOSSIL ELK.

Engraved by T.L. Busby.

Pub.d by J. Stratford, June 1.1811.

PLATE 72

ZOOLOGY.

Genus—CERVUS. *Species*—CERVUS FOSSILIS.

AMONGST the singular fossil remains which have at different times been excavated from the earth, no object has so much attracted the curiosity and attention of naturalists as the horns of the Elk, found at very considerable depths. It is well known that they are found only in the bogs and morasses of Ireland, the other bones of this curious animal incognitum, are seldom found existing in the earth united to each other or to the horns. The skull indeed is sometimes preserved very perfect, firmly adhering, but giving us no clear idea of the size of the creature which could have carried such an enormous weight. Of the different shape of the horns of the two Elks now known to exist, and to have survived the changes which the world has evdiently undergone, a particular account is given by Mr. PENNANT in his Arctic Zoology, with an elegant and exact delineation by Mr. STUBBS, and to which we refer our Readers.

The Horns which we have at present to describe, are excessively large, being twelve feet in the girth; the head one foot nine inches in length, into which the horns are fixed very firmly, and which are in a most perfect fossil state. It is no doubt the same which having been found in a bog at a considerable depth in Ireland in early times, was transmitted as a present to King Henry the Eighth. Since that time it has remained in the hands of royalty; afterwards as a gift from King William the Third to the first Duke of Portland, and since that disposed of to an eminent dealer in curiosities of the name of PACEY.

Such is the modern history of this surprising natural production, as it has been found so deeply covered with earth, forty or fifty feet at least, we must consider it as an antidiluvian, and that it has become extinct, (perhaps at that period) is evident, since no animal has so great a width of horns as the above, in any known part of the globe; and though frequently found in Ireland, is discoverable no where else. An Irish gentleman being asked for an explanation of the phenomena, very gravely answered, " that he thought it not a natural but an accidental cause, for that his horns where much too wide and spreading to admit of passing into the door of Noah's ark." That the earth has been universally deluged, or only partially covered, has formed an object of investigation with the greatest philosophers and naturalists, nor are they yet all agreed as to what was the proximate cause of such an amazing though temporary change; some, like Whiston, supposing it originated from the near approach and recession of a comet; others to an alteration in the inclination of the earth's axis, such a change, whether sudden or gradual, might influence the tides in a very powerful manner, and continents of land, the qualities of different animals be wholly destroyed, and that which form a part of the sea might have once existed as solid land.

OF THE CAMEL TRIBE.

ALL the animals of the Camel tribe are mild and gentle in their disposition. In a wild or native state they are not caught without great difficulty, yet when taken young, and trained to labour, they are made very serviceable to mankind. There are seven species, two of which only are found on the old continent, the rest being confined to the Alpine countries of Chili and Peru. It is supposed that most, if not all of them are gregracious, associating together in vast herds. The females have two teats, and seldom produce more than one young one at a birth. The hair of these animals is of a soft and silky texture; and their flesh forms a very palatable food.

In the lower jaw of the Camels there are six front-teeth, which are somewhat thin and broad. The canine-teeth are at a little distance both from these and the grinders: in the upper jaw there are three, and in the lower two. The upper lip is cleft or divided.

These animals like all the other genera of their order, are furnished with four stomachs, in consequence of which they not only live solely on vegetable food, but ruminate or chew the cud. They swallow their food unmasticated. This is received into the first stomach, where it remains some time to macerate; and afterwards, when the animal is at rest, by a peculiar action of the muscles, it is returned to the month in smll quantities, chewed more fully, and then swallowed a second time for disgestion.

M m

OF THE ARABIAN CAMEL.

This species is chiefly found, in a wild state, in the deserts of Arabia and Africa, and in the temperate parts of Asia. It is that, with a single hunch on its back, which we so frequenty see exhibited in the streets of this country. In many parts of the east it is domesticated, and in carrying heavy burdens over the sandy deserts, supplies a place which the horse would not be able to fill. The tough and spungy feet of these animals are peculiarly adapted to the hot climates, for, in the most fatiguing journies, they are never found to crack. The sand seems indeed their element, for no sooner do they quit it, and touch the mud than they can scarcely keep upon their feet, and their constant stumbling in such situations is exceedingly dangerous to the rider. Their great powers of abstaining from drinking, enables them to pass unwatered tracts of country for seven, eight, or, as Leo Africanus says, for even fifteen days, without requiring any liquid. They can discover by their scent at half a league's distance, and, after a long abstinence, will hasten towards it, long before their drivers perceive where it lies. Their patience under hunger is such that they will travel many days fed only with a few dates, or some small balls of barley-meal; or on the miserable thorny plants they met with in the deserts. M. Denon informs us, that during his travels in Egypt the Camels of the Caravan had nothing in the day but a single feed of beans, which they chewed for the remainder of the time, either on the journey, or lying down on the scorching sand, without exhibiting the slightest signal of discontent.

A large Camel will bear a load of a thousand or twelve hundred pounds, and with this it will traverse the deserts.

When about to be loaded, at the command of the conductor, the animals instantly bend their knees. If any disobey, they are immediately struck with a stick, or their necks are pulled down; and then, as if constrained, and uttering their groan of complaint, they bend themselves, put their bellies on the earth, and remain in this posture till they are loaded and desired to rise. This is the orgin of those large callosities on the parts of their bellies, limbs, and knees, which rest on the ground. If over-burdened, they give repeated blows with their heads to the person who oppresses them, and sometimes utter the most lamentable cries.

They have a very great share of intelligence; and the Arabs assert that they are so extremely sensible of injustice and ill-treatment, that when this is carried too far, the inflictor will not find it easy to escape their vengeance; and that they will retain the remembrance of an injury till an opportunity offers for gratifying their revenge. Eager, however, to express their resentment. they no longer retain any rancour, when once they are satisfied; and it is even sufficient for them to *believe* they have satisfied their vengeance. Accordingly, when an Arab has excited the rage of a Camel, he throws down his garments in some place near which the animal is to pass, and disposes them in such a manner that they appear to cover a man sleeping under them. The animal recognizes the cloaths, seizes them in his teeth, shakes them with violence, and tramples on them in a rage. When his anger is appeased, he leaves them, and then the owner of the garments may make his appearance without any fear, load, and guide him as he pleases. "I have sometimes seen them, (says M. Sonnini,) weary of the impatience of their riders, stop short, turn round their long necks to bite them, and utter cries of rage. In these circumstanbes the man must be careful not to alight, as he would infallibly be torn to pieces: he must also refain from striking the beast, as that would but increase his fury. Nothing can be done but to have patience, and appease the animal by patting him with the hand, (which frequently requires some time,) when he will resume his way and his pace of himself."—Like the Elephant, Camels have their periodical fits of rage, and during these they sometimes have been known to take up a man in their teeth, throw him on the ground, and trample him under their feet.

There is no mode of conveyance so cheap and expeditious as that by Camels. The merchants and other passengers unite in a caravan to prevent the insults and robberies of the Arabs. These caravans are often very numerous, and are always composed of more Camels than men. In these commerical travels their march is not hastened : as the route is often seven or eight hundred leagues, their motions and journies are regulated accordingly. The Camels only walk and travel thus from ten to twelve leagues a day. Every night they are unloaded, and allowed to pasture at freedom.

When in a rich country, or fertile meadow, they eat, in less than an hour, as much as serves them to ruminate the whole night, and nourish them during the next day. But they seldom meet with such pastures, neither is this delicate food necessary for them. They seem to prefer wormwood, thistles, nettles, broom, cassia, and other prickly vegetables, to the softest herbage. As long as they find plants or brouze, they easily dispense with water. This faculty of obstaining long from drink proceeds not, however from habit alone, but is an effect of their structure. Till very lately the Camels have been supposed to possess independantly of the four stomachs common to ruminating animals, a filth bag, which served them as a reservoir for holding water. From a preparation, however, in the collection of Mr. John Hunter, it appears that this filth bag never existed but in idea. The second stomach is of peculiar construction, being formed of numerous cells several inches deep, having their mouths uppermost, and the orifices apparently capable of muscular contraction. When the animal drinks it probably has a power of directing the water into these cells, instead of letting it pass into the first stomach and when these are filled the rest of the water will go into that stomach. In this manner a quantity of water may be kept seperate from the food, serving occasionally to moisten it in its passage to the true stomach, for several day.

When travellers find themselves much in want of water, it is no uncommon thing to kill a camel for what he contains, which is always sweet and wholesome.—Aristotle says, that the Camel always disturbs the water with its feet before it drinks : if this be the case which, it must be confessed, seems very doubtful, it is done to chase away the almost innumerable swarms of insects with which the waters of warm climates abound.

"Of all animals (says Comte de Buffon) that man has subjugated, the Camels are the most abject slaves. With incredible patience and submission they traverse the burning sands of Africa and Arabia, carrying burthens of amazing weight. The Arabians consider the camel as a gift sent from heaven, a sacred animal, without whose assistance they could neither subsist, traffic, nor travel. The milk of the Camel is there common food. They also eat its flesh ; and of its hair they make garments. In possession of their Camels, the Arabs want nothing, and have nothing to fear. In one day they can perform a journey of fifty leagues into the desart, which cuts off every approach from their enemies. All the armies in the world would perish in pursuit of a troop of Arabs. By the assistance of the Camel, an Arab surmounts all the difficulties of a country which is neither covered with verdure, nor supplied with water. Notwithstanding the vigilance of his neighbours, and the superiority of their strength, he eludes their pursuit, and carries off with impunity all that he ravages from them. When about to undertake a depredatory expedition, an Arab makes his Camels carry both his and their own provisions. When he reaches the confines of the desart, he robs the first passengers who come in his way, pillages the solitary houses, loads his Camels with the booty, and, if pursued, he accelerates his retreat. On these occasions he displays his own talents as well as those of the animals. He mounts one of the fleetest, conducts the troop, and obliges them to travel day and night, without almost either stopping, eating, or drinking, and, in this manner, he often performs a journey of three hundred leagues in eight days.''

With a view to his predatory expeditions, the Arab instructs, rears, and exercises his Camels. A few days after their birth he folds their limbs under their belly, forces them to remain on the ground, and in this situation loads them with a tolerably heavy weight, which is never removed but for the purpose of replacing it by a greater. Instead of allowing them to feed at pleasure, and drink when they are thirsty, he begins with regulating their meals, and makes them gradually travel long journies, diminishing at the same time the quantity of their aliment. When they acquire some strength they are trained to the course, and their emulation is excited by the example of horses, which, in time, renders them not only fleet, but more robust than they would otherwise be. In

Egypt thair value is, according to their goodness, from two to five hundred livres.

The saddle used by the Arabs is hollowed in the middle, and has at each bow a piece of wood placed upright, or sometimes horizontally, by which the rider keeps himself on his seat. This, with a long pocket to hold provisions for himself and his beast, a skin of water for the rider, (the animal being otherwise well supplied) and a leather thong, are the whole of the equipage that the Arab traveller stands in need of, and with nothing more than these he is able to cross the desarts.

The pace of the Camel being a high trot, M. DENON says that when he was first mounted on one of these animals, he was greatly alarmed lest this swinging motion would have thrown him over its head; he, however, was soon undeceived, for on being fixed in the saddle, he found that he had only to give way to the motion of the beast, and then it was impossible to be more pleasantly seated for a long journey, especially as no attention was requisite to guide the animal, except in making him deviate from his proper direction. " It was," said he, " entertaining enough to see us mount our beasts; the Camel who is so deliberate in all his actions, as soon as his rider leans on his saddle, preparatory to mounting, raises very briskly, first on his hind, and then on his fore legs, thus throwing the rider first forward and then backward; and it is, not till the fourth motion that the animal is entirely erect, and the rider finds himself firm in his seat. None of us were able for a long time to resist the first shake, and we had each to laugh at his companions."

When the traveller is not in haste, or when he accompanies a caravan, the progress of which is always slow on account of the Camels of burthen, a kind of covered litter is fixed on one of these animals, in which he is tolerably at his ease, and where he may even sleep if he chuses.

The drivers of the loaded Camels have each a stick, which they use sparingly, if occasion requires; and those who ride, whip their animals with a long strap of leather, at the same time urging them with a clicking of the tongue, the same as the Europeans use to their Horses. It has been asserted by Mr. PENNANT and some other writers, that Camels, may be made to go more freely by whistling to them; this, however, is a mistake; and the Bedouin Arabs, who own immense numbers of Camels, not only never whistle themselves, but it even gives them pain to hear others whistle.

The flesh of the Camel is dry and hard, but not unpalatable. It is so much esteemed by the inhabitants of Egypt, that in Cairo and Alexandria, it was, not long ago, forbidden to be sold to the Christians. In Barbary, the tongues are salted and smoked for exportation to Italy and other countries, and they form a very good dish. The hair is an important article of commerce, serving for the fabrication of the tents and carpets of the Arabs; and leather is made of the skin. In the materica medica of China, the different parts of the Camel occupy a conspicuous place: the fat is called the oil of bunches, and the flesh, the milk, the hair, and even their dung, are admitted into the prescriptions of the Chinese physicians.

[This page was blank in the original]

Drawn & Engrav'd by T.L.Busby.

KANGUROO.

Pub. by J. Stratford, July, 1ˢᵗ 1811.

PLATE 73

[492]

ZOOLOGY.

Genus—DIPUS, *or* JERBOA.
Species—DIPUS TRIDACTYLUS, *or* KANGAROO.

(From Mr. Polito's Museum.)

Generic Character.—Incisores, or cutting teeth of an irregular number, the lower ones placed horizontally, a wide open space standing between the incisores and grinders. The hind legs very long, generally three times as long as the fore legs, ears rounded and projecting, body containing an outward pouch for the young, placed like an apron across the body.

IN a former number of the ARCANA, we have imparted the Dipus Muscola, a small animal, evidently of the Kangaroo species, similar to the present, and existing only at Botany Bay and Van Diemens Island. It was first discovered during the voyage of Captain Cook, and exhibits to the human mind, as strange an assemblage of forms as the most eccentric fancy can conceive. Its head and neck are small and taper, resembling those of a Squirrel; the nose of a dark brown, adorned with whiskers; the neck and fore legs very short, but the body extended like a bag of shot, for four or five feet, with enormously strong hind legs, which are calculated like those of a flea, by a particular spring in their motion, to leap up to a considerable distance. The tail is also so thick and strong that it frequently rests upon it as if upon a chair, and steadying its upright direction, by its various expansion of the tail. The age of the Kangaroo is not exactly known, but is supposed to be about twenty-five years, and those which have been deposited in the Queen's Palace at

ZOOLOGY.

Kew have been as healthy, or even more so than in their native country, bringing forth one or two young Kangaroos at a birth, which they carry in a pouch or apron, where the little creatures may be seen peeping out of the bag, whenever the mother directs herself to sitting up. At first they are very small and tender in their limbs, requiring the fostering warmth and affection of the mother, but, in four or five months, their growth rapidly encreases, and they gradually desert their maternal habitation. The Kangaroo is eaten both by the Europeans and the natives. When in a broiled state, it is a pleasant and agreeable food, and said to taste like a Hare. We conceive that it is not much unlike the Hare in its general character, as it is calculated most for a tame fearful animal, being without all means of offence. Nature has kindly provided the path for its safety in the swiftness and extent of its leaps, which are reported to be from rock to rock, for twenty feet at a time, over bush and brier. When the Kangaroo threatens to fight with its mates, it is said to lean forward upon its fore feet, and to use its long hind toes in a kicking posture, like the Ass. Its front teeth being placed flat in the mouth, seem not to be calculated for a severe and cruel bite ; we may, therefore, conclude that its general character is placable and serene ; that it is placed by Providence, in a situation where no larger animal can injure or destroy it, to excite our wonder and strong perception of the variety of Nature.

> " He reigns the monarch of the pathless waste,
> " Where solemn silence spreads her aweful scene !
> " Save where the savage strikes the finny tribes,
> " Or scanty worm drawn from the coral rock ;
> " While screaming Penguins beat the sounding shore,
> " And the dark thunder, from the mountain pours
> " A heavy flood on the defenceless wretch,
> " All houseless, homeless, as he wanders on,
> " To seek some cave to shield him from the storm !"

[This page was blank in the original]

Drawn by O. Perry.

Engraved by T.L. Busby.

Pub. by J. Stratford. July. 1. 1811.

PHALŒNA FENESTRA.

PLATE 74

[496]

ZOOLOGY.

Distinctive Genus—PHALÆNA. *Division*—ARCUATA. *Species*—PHALÆNA FENESTRA.

THE Genus Phalæna, although not possessing the vivid tints of the Butterfly tribes, yet powerfully arrests our attention by the variety of its texture and the glossy talc or transparent membranes of the different parts. This plainness of colouring becomes, however, in the hands of Nature subservient to an elegance of shape which may vie with all the objects of Nature.

The present Moth is distinguished by the richness of its border, and two quadrangular spots in the upper wings: these are transparent, and of a greyish brown shade : the edges and end of the upper wing are arcuated with brown indented lines, similar to those formerly given to the Atlas Moth, excepting that the present is not quite so large. The under wings in this Moth are adorned with spots, similar to those of a Peacock's tail; and the middle and border of them is of a brown colour, mixed with veins of yellow. As this Moth has not hitherto been named by Linnæus, Fabricius, or any other writer, I have ventured to give it the name of Fenestra, from the circumstance of the quadrangular transparency of a diamond form in the upper. Our Readers, by turning back to the two species of Fulgora mentioned in a former Number, will perceive the difference of form in these Insects, as to the legs, and also the proboscis and separation of the wings. In the Papilio or Butterfly he will find that there is an evident knob or swelling at the end of the Antennæ, which will distinguish it also from the Sphynx, each of which we have endeavoured to elucidate in a former Number. The number of Microscopic Moths, only visible by means of a glass, a- mounts, it is said, to three thousand. If this be the case in so small an instance, we must despair ever of enumerating the remaining small Genera ; and it seems very remarkable that the small Genera of Papilios should be much less nume- rous in all the regions at present known. New Holland and the adjacent islands in the South Seas will most probably fur- nish us with a new Encyclopædia of Insects and Birds, too numerous to form any exact detail, and to which we must al- ways be adding. So boundless, so extensive are the works of the Creator, and so strong is the impulse of human curiosity to explore the relationship of things, that we may exclaim with the Roman poet Virgil :

"Felix qui potuit rerum cognoscere causas."

[This page was blank in the original]

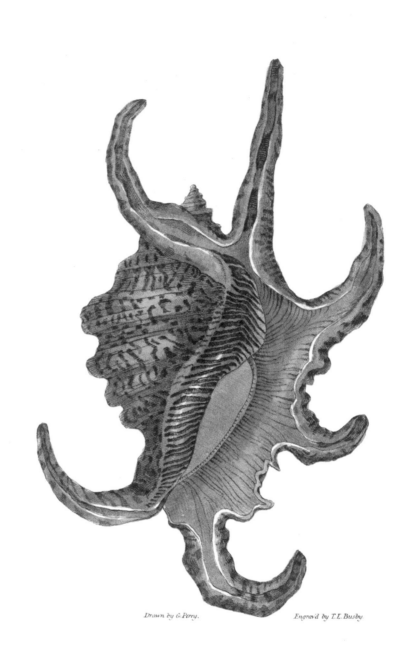

Drawn by G. Perry.   Engrav'd by T.L. Busby.

STROMBUS DIVERGENS.

Pub'd by J. Stratford, July, 1st 1811.

PLATE 75

[500]

ZOOLOGY.

Genus—STROMBUS.
Species—STROMBUS NIGRICANS.

Generic Character.—Shell univalve, spiral, the cheek or Maxilla Oris, turned backwards and spread out like a flap, the cheek also cut open and furrowed at the top and bottom, near to where it joins the beak; the beak twisted inwards and backwards.

NONE of the shells which we have already presented to our Readears, possess a greater singularity than the present univalve, commonly called the Strombus Nigricans. Having in a former Number described the Strombus Solitaries, which is distinguished only by one horn, we now proceed to describe the second shell of that genus, in which the Reader will easily perceive a general similitude. The general colour of this shell is of a yellowish brown, shaped into painted lines, in the inside of the mouth. There are also other lines of a black and some of an orange colour, which are variously diversified, and add very much to the richness and splendour of the whole. There are also five principal horns, placed upon different parts of the shell : these are streaked with brown and black, especially at the top of each, and bears a considerable resemblance to the horns of a Cow. The spire is short, as in the cone, in which respect it resembles also the Voluta Genus; the beak is generally twisted backwards.

In different sea counties of England, as Cornwall, Devonshire, and Hampshire, are found, three or four different fossil shells of the Strumbus kind, and of the same genus as above described; these, however, are different in their particular specific forms, and must therefore have been containing animals of a race now extinct. In the British Museum there is a beautiful preserved collection of the fossil shells of Hampshire, which exhibit a delicate form of ornaments, resembling the Herculaneum Vases, but having lost all their original colours, by the lapse of ages, have received a brown tinge from the surrounding earths, some of them however are white and have kept their exquisite polish.

We are much indebted to Mr. Brander of Hampshire, who generously imparted it to the British Museum, and employed the learned Dr. Solander to edit a small treatise upon these remains. Thus nobly appropriating the advantages of fortune to the improvement of science and knowledge, how noble does the human mind appear ! Nor can we here omit the princely collection of shells formed by the gift and legacy of Mr. Cracherode, which has not, nor ever will be rivalled, containing specimens which cannot be removed nor injured. Owing to the judicious arrangement of the British Trustees, it will be a standing monument of the general taste of the English Nation; nor will it fail to excite others, in after times, to similar exertions ; and other Sloane's and Linnæus's will illuminate the regions of Natural History, with successive splendour.

[This page was blank in the original]

ZEBRA.

Drawn & Engraved by T.L.Busby, from Palace Menagerie.

Pub.d by J. Stratford, 1 July, 1810.

PLATE 76

ZOOLOGY.

Genus—EQUUS.
Species—EQUUS ZEBRA.

THE Genus Equus or the Horse has been separated by the most eminent modern Naturalists, into five distinct families of animals, each of which is considered as a distinct species of Horse. First, the Arabian Horse, a native of Africa and Asia; secondly, the Tartarian Horse, a native of the kingdom of Tibet only, and nearly extinct; thirdly, the Quagga, a native of Caffraria in Africa; fourthly, the Zebra, of the same country; and, fifthly and lastly, the Ass, a native of Judea and Egypt.

The Zebra, somewhat resembling the Mule in its form and size, has a large head and ears. Its body is round and plump; its legs are delicately small, and of course rather unlikely to carry any great weight, and of which it always seems naturally impatient. The skin is as smooth as satin, and adorned with elegant stripes, like ribbands, which in the male are brown on a yellowish white ground, and in the female black on a white ground. Zebras inhabit the scorching plains of Africa, their vast herds, affording an agreeable relief to the weary eye of the traveller. They assemble in the day on the extensive plains of the interior of the country, and by their beauty and liveliness adorn and animate the dreary scene. All the attempts which have been made to tame this animal have been in vain; to render it serviceable to mankind has been hitherto fruitless. Wild and independent by nature, it seems ill adapted to servitude and restraint. If, however, it were taken young, and much care was bestowed on its education, it might probably be, in a great measure, domesticated. A beautiful male **Zebra** at Exeter 'Change, London, which was burnt to death by the mischievous act of a Monkey setting fire to the straw on which he lay, appeared to have entirely lost his native wildness, and was so gentle, as to suffer a child of six years old to sit quietly on his back, without exhibiting the least sign of displeasure; he was familiar even with strangers, and received those caresses that are given to the Horse with evident satisfaction. The one, however, that was some years ago kept at Kew seemed of a more irreconcileable nature; no one dared to approach, except the person who was accustomed to feed it, and who alone could mount on its back. Mr. Edwards saw this animal eat a large paper of tobacco, paper and all, and was told it would eat flesh and any kind of food that was given it. This, however, might proceed from habit or necessity, in its long voyage to this country; for it is supposed, that in its native country, these animals all feed, like Horses and Asses, on the vegetable kingdoms.

In the vallies and provinces near to the Cape of Good Hope there are herds of Zebras, and a penalty of six dollars is inflicted on any person who shoots one of them. When any of the young ones happen to be caught alive in the woods, they are always taken to the town and ordered to be sent to the Governor.

The present Engraving is executed from a specimen in the Museum of Mr. POLITO, Exeter 'Change.

Remarks on the Natural History of the BUFFALO.

IT has been supposed by certain Naturalists, who delight in difficult theories, that the Bison, the Cow, and Buffalo, are only different races of the same original animal, which have been altered in their nature and qualities by the transposition of climate, and variety of sustenance. Buffon seems first to have adopted this strange assimilation, which does not appear justified by either their habits or formation. The Buffalo is a native of India and Africa, and has a great resemblance to a large English Ox, but it differs much in its horns and in some particulars of its internal structure. It is larger than the Ox ; the head also bigger in proportion, the forehead higher, and the muzzle longer. The horns are large, of a compressed form with the exterior edge sharp; they are straight for a considerable length from their base and then bend slightly upwards. The general colour of the animal is black, except the forehead and tip of the tail, which are of a dusky white ; the hunch is not, as many have supposed it, a large fleshy lump, but is occasioned by the bones that form the withers, being continued to a greater length than in most other animals.

Buffalos have been introduced into Italy, and are naturalized and domesticated there ; they constitute a great part both of the riches and food of the commonalty. They are employed in Agriculture ; and butter and cheese are made from their milk: they are fond of swimming in rivers, and will dive ten or twelve feet deep to force up with their horns the aquatic plants, which they eat while swimming.

In the Eastern Countries also, the Buffaloes are domesticated. It is said to be a curious prospect to see, morning and evening, large herds of them cross the Tigris and Euphrates. They proceed all wedged against each other, the herdsman riding on one of them, now and then standing upright, and sometimes couching down, and if any of the exterior are out of order, stepping lightly from back to back, to drive them along.

The skin and horns of the Buffalo are its most valuable parts ; the former being extremely strong and durable, consequently well adapted for various purposes in which the strongest leather is wanted. The horns have a close grain, are strong, and bear a good polish, being on that account much valued by cutlers and other artificers. The flesh is said to be excellent, and it is so free from any disagreable taste as to resemble beef as nearly as possible.

Very different from these is the wild Buffalo of the Cape of Good Hope, distinguished by Linnæus, as the Bos Cafer. It is a very large and enormously strong animal. The fore parts of the body are covered with long coarse and black hair. The horns are thick and rugged at the base, sometimes measuring three feet in length, and lying so flat upon the head as almost to cover it. The ears are large and slouching, the body and limbs thickly muscular, bearing a most fierce and malevolent aspect. In the Plains of Caffraria they are so common, that it is by no means unusual to see a hundred and fifty or two hundred in a herd. They generally retire to the woods in the day time, and at night go out into the plains to graze.

Treacherous in the extreme, they frequently conceal themselves among the bushes and trees in the securest coverts, and lurk in secret, until some unfortunate traveller comes past, when the animal at once rushes out into the road ; and the traveller has but a small chance of escaping, except by climbing into a tree, if he is fortunate enough to be near one. Flight is of no avail; he is speedily overtaken by the furious beast, who, not contented with throwing him down and killing him, stands over him for a long time afterwards, trampling him with his hoofs, and crushing him with his knees; and not only mangles and tears the body to pieces with his horns and teeth, but likewise strips off the skin, by tearing it with his tongue. Nor does he perform all this at once; but often retires to some distance from the body, and returns with savage ferocity to gratify afresh his cruel inclinations.

As professor Thunberg was travelling in Caffraria, he and his companions had just entered a wood, when they discovered a large old male Buffalo, lying quite alone, in a spot that for a few square yards was free from bushes. The animal no sooner perceived the guide, who went before, than, with a horrible roar, he rushed upon him. The man turned his horse short round behind a large tree, and the Buffalo run forwards to the next passenger, goring his horse so dreadfully in the belly that he died soon after. These two climbed up into the trees, and the furious animal made his way towards the rest, of whom the professor was one, who were approaching, but at some distance. A Horse, without a rider, was in the front, and as soon as the Buffalo saw him, he became more outrageous than before, and attacked him with such fury, that he not only drove his horns through the Horse's breast, but even out again at the very saddle. This Horse was thrown to the ground with such excessive violence, that he instantly died, and many of his bones were broken; the professor had just time to witness this last circumstance, and to take refuge also in a tree, and the Buffalo turning suddenly round soon gallopped away. Some time after this, the professor and his party espied an extremely large herd of Buffalos grazing on a plain. Being now sufficiently apprized of the disposition of these animals, but knowing that they would not attack any body in the plain, they approached within forty paces and fired amongst them. The whole troop, notwithstanding their intrepidity, surprized by the sudden flash and report, turned about and made off towards the woods; the wounded Buffalos separated from the rest of the herd, by their inability to keep pace with them; amongst these was a particular Bull, which ran with fury at the party, but fell exhausted at their feet.

The Caffrarians hunt the Buffalos with arrows and javelins, which they hurl with great dexterity. When a Caffre has discovered the place where several Buffalos are collected together, he blows a pipe, made of the thigh bone of a Sheep, which is heard at a great distance. The moment his comrades hear this notice, they run up to the spot and surrounding the animals, which they take care to approach by degrees, lest they should alarm them, throw their javelins at them. This is done with so sure an aim, that, out of eight or twelve, it is very seldom that one escapes.

It sometimes happens, however, that while the Buffalos are running off, some one of the hunters, who stands in the way, is tossed and killed; but this is a circumstance not much regarded by the Caffrarians. When the chace is ended, each Hottentot cuts off and takes away his own share of the game. Certain European hunters once chaced a Buffalo in the neighbourhood of the Cape, and having driven him into a narrow place, he turned round and instantly pushed at one of his pursuers, who had on a red waistcoat, a colour which they detest; the poor man, to save his live, ran to the water, instantly jumped in, and swam off; the fierce animal followed the man so closely, that he had no opportunity of escape, except diving; he dipped over head, and the Buffalo loosing sight of him, swam on towards the opposite shore, which it intended to have reached, had it not been killed by a gun from a ship, lying at a little distance; the skin was presented to the governor of the Cape, who placed it in his museum.

Like the Hog, this animal is fond of wallowing in the mire; the flesh is lean, but juicy and of a rich flavour. The hide is so thick and tough that a musquet proof covering is formed from it: even while the animals are alive, they are said to be impenetrable to a musquet ball. These have generally a mixture of tin infused for the purpose of hardening them, yet even these are flattened sometimes, and drop off harmless, so strong and defensible is their armour.

We shall now proceed to describe the Bison, or American Cow, which is peculiar to North and South America, and has, in its general form, a considerable resemblance to the Buffalo, but is considerably smaller, yet very strong and resolute in its own defence. The fore parts are excessively thick and broad, but the hinder parts are comparatively very slender.

These animals range in droves, feeding in the open Savannahs, morning and evening, retiring during the sultry times of the day to rest near shady rivulets and streams of water. They frequently leave so deep an impression of their feet on bog-land, (from the great weight of their bodies,) as to be quickly traced down by the artful hunters. In this undertaking it is necessary that the men should be particulaaly careful, since, when slightly wounded, the animals become excessively furious. The hunters pursue them against the wind, as the faculty of smell, in the Bisons, is so exquisite, that the moment they perceive the scent of their enemy, they retire with the utmost precipitation. With a propitious and favourable wind they approach very near to them ; since the animals are frequently almost blinded by the hair that covers their eyes, they then aim their piece at the shoulder, by which they generally bring them down at one shot ; but if they do not fall, they instantly run upon their enemy, and with their horns and hoofs, as their natural weapons, tear him in pieces and trample him to the earth. They are so amazingly strong, in flying through the woods from their pursuers, that they frequently brush down trees as thick as a man's arm ; and, although the snow should be ever so deeply drifted, they plunge through it with all the ease imaginable.

In Canada, the hunting of the Bison is a very general employment to the natives. Their method is to draw up in a large square, or phalanx, which they gradually diminish, setting fire to the grass and trees. As the fire goes on they advance their guard, closing up their ranks as they proceed ; and the animals, alarmed by the light, gallop confusedly about untill they are hemmed in so close, that not one can escape from its pursuers.

In Louisiana the men mount on horseback, each with a sharp crescent pointed spear in his hand; they approach with the wind, and as soon as the animals smell them they instantly make off; but the sight of the horses moderates their fear, and the majority of them are at a certain time of the year so fat and unweildy, as readily to be enticed to slacken their pace. As soon as the men overtake them, they endeavour to strike the crescent just above the ham, in such a manner as to cut through the tendons, and render them afterwards an easy prey.

The sagacity with which these animals defend themselves from wolves is remarkable. When they scent a drove of these ravenous creatures, the herd throws itself into the form of a circle, having the weakest in the middle, and the strongest placed on the outside; thus presenting a formidable front of horns.

Attempts have been made to domesticate them by catching the calves and herding them with the common kind, in hopes of improving the breed. This has not, however, been found to answer; for when they grew up, they always became impatient of restraint, and from their great strength would break down the strongest inclosures, and entice the tame cattle to follow them.

The uses of the Bison are various. Powder flasks are made from their horns. The skin forms an excellent buff leather, and when dressed with the hair serves the Indians for cloaths and shoes. The Europeans of Louisiana use them for blankets, finding them light, warm, and soft. The flesh is a considerable article of food, and the bunch on the shoulders is esteemed a great delicacy. The bulls, when fat, frequently yield one hundred and fifty pounds weight of tallow, which forms a considerable article of commerce. The hair or wool is spun into hair stockings or garters, that are very strong, and look as well as those made of the finest sheeps wool. The only circumstance which they seem to be deficient in, is their beauty, in which Nature seems entirely to have deserted them, giving them a large pair of staring eyes, half covered with a long flowing mane, like that of a Lion, but which it does not resemble in its solitary life. Its faults of temper are fully compensated by its good qualities, and Man, the Lord of the Creation, who can kill or make alive, can hardly be said to enjoy so much perfect liberty, or unbounded pleasure as the Bison, which ceases under a sudden consciousness of death, which can best make his usefulness apparent.

There is a singular and affecting trait in the character of the Bison, when a calf. Whenever a cow Bison falls before the murdering hand of the hunters, and happens to have a calf, the helpless young one stays by its fallen dam, with signs expressive of strong and natural affection. The dam thus secured, the hunter makes no attempt on the calf, (knowing it to be unnecessary,) but proceeds to cut up the carcase: then laying it on his Horse, he returns towards home, followed by the poor calf, thus instinctively attending the remains of its dam.

[This page was blank in the original]

ÆGYPTIAN RAM.

Drawn & Engrav'd by F.L. Busby, from Polar's Menagerie

Pub'd by J. Stratford, Aug' 1st 1811.

PLATE 77

Genus—OVIS ARIES, *Anglicè*, RAM or SHEEP.

Generic Character.—The Male armed with crooked horns, the Female generally without them, (or else very small); the feet cloven and pointed; canine teeth none; disposition harmless; back covered with curled wool.

THE great variety of the horns in the Sheep tribes, has very much attracted the attention of Natural Historians, and may serve, in many respects, to distinguish the different species of various countries. The length or shortness of the legs, also form a striking characteristic of many kinds, as the Icelandic, Scythian, Pomeranian, and others.

The Sheep, or Ram, which we now present to our Readers is, perhaps, one of the most remarkable, and scarcely ever seen in England; it is exactly delineated from a living specimen in Mr. Polito's Museum; it bears in the character of its horn, the form of the ancient representation of Jupiter Ammon. From this circumstance, the Cornu Ammonis was placed, by the ancient Lybians, in the temple of their favourite deity, and was afterwards visited in the city of Oasis, by Alexander the Great. Probable conjectures would lead us to suppose this animal to be one of the twelve signs of the Zodiac. It appears as if the skeleton of this Ram, with a complete head and horns, had been described by two naturalists, Buffon and Pennant, who never could obtain the original animal alive, but have denominated it, from its form, Caprea Capricorna. In several of the coins of ancient Macedonia, Alexander impressed the horn upon the image

P p

of his own head.   As he had a specious pretence that he was originally descended from Jupiter Ammon, this superstitious idea lulled the envy of his great power and conquest in the minds of all his Pagan subjects.

The Aries Ammonis is rough and hairy on the back, breast, and legs; it has a grave and dignified deportment, and is said to be only found in the higher mountains of Egypt; like to the Ibex, which is now very rare, in the Alpine summits of Switzerland.   His legs are long, and of an irregular shape; his colour resembling a coffee colour mixed with a paler brown; and, what is rather extraordinary, the ears are spread out horizontally through the middle of the horn; his disposition seems to be good-natured and placable, not at all irascible, like the common Goat, which it much exceeds in size.

Few animals render greater or more essential services to mankind than the Sheep ; they supply us both with food and cloathing.   The wool alone of the common Sheep supplies some countries with an astonishing source of industry and wealth.   It is very probable, however, that the Merino wool, the breed of which has been so largely imported by government, may, in a few years, wholly degenerate from its shortness and silky quality, by means of the coldness of the winters of Great Britain, which always lengthens and makes coarse the hair of all other animals; and in their form and gait they are generally considered of too small a race to improve the English flocks.

[This page was blank in the original]

CROWN CRANE.

Pub.d by J. Stratford, Aug.t 1. 1811.

PLATE 78

Genus—ARDEA. Species—ARDEA CORONATA, or CROWNED CRANE.

Generic Character.—Bill partly cylindrical, the thighs feathered only half way to the knees, the legs long and formed for wading in the water; the tail short and flattened, the toes more or less webbed in the different species, three before and one behind, placed as a prop in walking, a circular or irregular leather-like flap, generally dividing the eye and bill.

THE superb bird, which forms so elegant an ornament to the Crane class, is a native of Africa, and not very unfrequently brought over alive to England, its grandest ornament is its crest or crown, which are expanded and spotted, like the quills of a Porcupine : on each side of the head there is a beautiful red membrane, invested with a dark blue crown, as a base, or foundation, from which the quills, or spines, arise in a beautifully radiated form. The neck and breast have exquisite hairy folds, green mixed with grey, curling and projecting. The body is a mixture of black, grey, and white, in such a manner as to instruct and delight the eye of a painter. Like to the Lady-bird it is very apt to stand upon one leg for a considerable time, or change them rapidly with a sort of dancing. They reside, like the Crane or the Heron, at the edge of marshy pools, and associate in some solitary plains in immensely numerous bodies, for the purposes of migration to distant regions, at different seasons of the year. This, however, is not constantly the case, as a solitary life, at certain times, gives equal pleasure, and also some advantages in the pursuit of their prey.

Their different journeys are generally performed in the night, but their loud screams betray their course, which has been happily described by one of our English poets.

> —— " Part loosely wing the region ; part, more wise,
> " In common rang'd in figure, wedge their way
> " Intelligent of seasons ; and set forth
> " Their airy caravan, high over seas
> " Flying, and over land with mutual wing
> " Easing their flight.—So steers the prudent Crane
> " Her annual voyage, borne on winds ; the air
> " Floats as they pass, fann'd with unnumber'd plumes."

The flight of the Crane is always supported uniformly, though it is marked by different inflexions ; and these variations have been observed to indicate the change of the weather, a sagacity that may well be allowed to a bird, which, by the vast height to which it soars, is enabled to perceive the distant alterations and motions in the atmosphere. The cries of the Crane, during the day, announce rain, and their noisy tumultuous screams forebode a storm. If, in a morning or evening, they rise upwards, and fly peacefully in a body, it is a sign of fine weather; but if they keep low or alight on the ground, this menaces a tempest. Holben observes, that they generally have a centinel to keep watch, while the others are feeding, and on notice of danger, the whole flock are instantly on the wing. Cranes are found in France, in the spring and autumn, but are, for the most part, merely passengers. We are informed, that in the last century, they have been shot in Lincolnshire and Cambridgeshire in vast flocks, but none of late have been met with. The flesh is black, tough, and bad. The Stork has always been considered as a strong emblem of filial affection, carrying his aged parent upon his back, from place to place, with the kindest gratitude and regard.

[This page was blank in the original]

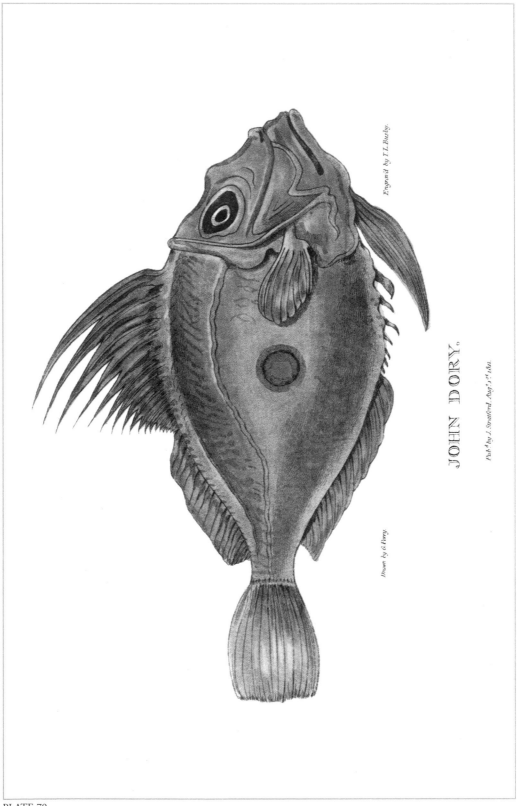

JOHN DORY.

Drawn by G. Perry.

Engrav'd by T. L. Busby.

Pub.d by J. Stratford, Aug.t.1.st.1811.

PLATE 79

Genus—ZEUS. *Species*—ZEUS FABER.

Generic Character.—The body flat, and back sharp, and erect, armed with spines, joined together by a slight finny substance. The underside of the body rounded like the upper one; the nose or trunk, in both the jaws, capable of a very wide extension; the teeth partly inclosed, and spotted in the centre of the body with a black spot on each side.

THE Zeus Faber, or John Dory, has been celebrated by epicures of the present and former times, as a fish of most delicious flavour; and the Romans are reported, by their historians, to have given almost any price for the largest. The flesh is not so firm, in its texture, as the Sole or Haddock, but resembles more the gelatinous qualities of Ling, which it a good deal resembles in its specific taste.

It is mentioned in the life of Mr. John Quin, the celebrated epicure, facetious companion, and most famous actor. In such luxurious subjects, he might be placed with Crassus the Roman, who paid three thousand sesterces, for a supper of fish, only composed for himself and six of his friends. In the comparison of the former with the Roman, it is difficult to say what our judgement is; Quin frequently went to Brighton and Essex, upon journeys which were rather out of his way of business, and mostly to such places, where the John Dory was plentiful. The Roman knight treated equally his friends and foes; the other had only a wish for being asked again, a much more selfish motive! To return, however, to the form of this fish, so singular in swimming straight, unlike the Sole or Plaise, described in a former Number, as having its eyes placed both on one side, and swimming in a flat direction, being mostly existing at the bottom; yet this forms a curious exception to that tribe, and has an eye on each side, and swims directly upright. The head is large, the body and tail small in comparison, the colour is brown and grey, mixed with a little portion of white; the pupil is beautiful and of dark blue, the iris yellow, the rest of the Dory is rather uncommon, by having a pearly appearance, when placed in a varying light; there is a spot of a black colour, placed in the centre on each side, and which occurs also in the Haddock. The superstitious Catholics suppose, that the finger and thumb of St. Peter, (or, as others imagine, of St. Anthony,) have impressed this mark on each side, so as to have remained on them ever since. It is fished for on almost all the coasts of Europe, and seldom exceeds in length sixteen inches, of which size it is generally found more solitary, and produces a price suited to the scarcity of the particular rarity of certain seasons. Its spines being so sharp, strong, and bent back, along with its great swiftness of motion, must constitute a safe defence against its submarine enemies.

[This page was blank in the original]

Drawn by G.Perry.   Engrav'd by T.L.Busby.

CONUS BANDATUS.

Pub.d by J. Stratford, Aug.t 1 1811.

PLATE 80

[528]

CONUS BANDATUS,

Generic Character.—Shell spiral oblong, the spire short; no beak; the mouth oblong and narrow, and ending at the base in an open channel rounded; the whole form of the shell having the shape of a cone, angular and pointed.

THE Cones consist of a family, or division of shell animals, rarely attaining to a larger size than six inches, but abounding in the minute species, of which we have observed, in the different museums of collectors, upwards of two hundred varieties, which require the microscope to examine their peculiar forms, and some of which have a fluted process situated on the columella, or central pillar. The Cone is is also one of the most compact of all other shells, and well defended from its external enemies in the ocean, by the narrow shape of its mouth. The spire is sometimes flattened at the top, as in the Cypræa, or Cowry, so as to be scarcely visible. The stripes are generally of different sizes and colours; the surface of a high and glossy polish.

Amongst the fossil specimens delineated in a former Arcana, No. 4, and discovered lately, near the environs of Paris, is a particular and elegant specimen of a Diluvian Cone, found eighty yards deep, in a bed of chalk. The top is curiously marked with circular bands; the colours in general lost, by being deposited in the earth, at the period of the great Deluge, and enveloped in its bleaching powers. Shells of Crabs, Spines of Snakes, and of Marine animals, have lately been found at Willsted and Brentford, in the county of Middlesex, as well as in the subterraneous tunnel, or passage, penetrating through Highgate Hill, Corallines and fragments of the Saw-fish were discovered. The teeth of Sharks, with glassy scoria, and a substance, much resembling burnt wood, were also mixed with a black resinous substance, yielding an agreeable smell, blazing in the fire with a yellow flame. The soil, in this beautiful part of Middlesex, is very hilly, compared with the other parts of the county, which are very flat, and is composed of lamina of very stiff clay, covering an unfathomable bed of lime of excellent quality. Thus the activity of Man has penetrated the Earth, the Sea, the Air. We find that new accidents display to Man, an immense field of natural knowledge, of which he has hitherto remained perfectly ignorant. Having investigated, in some degree, in our former Numbers, the general history of Fossils, we purpose, in a future part of our Work, to enter into a more particular and exact drawing and description of those circumstances, which may be termed novel, avoiding all unnecessary theory, and adhering to facts, which have come under the observation of the artist and naturalist, exhibiting the connections and analogies of some of the principal phœnomena of the Fossils of our globe.

A Dissertation on the CORVUS *Tribe of Birds of the* CROW *Kind.*

THE Corvus or Crow family are generally dispersed in all parts of the world, some of them being always found in each Climate hitherto known. They are numerous and very clamorous, attaching themselves by large flocks, one of which generally keeps guard, by being pearched upon some high tree or barn to give notice of danger to the rest by a particular scream, which to surprize us still more is answered by a different note and a rapid decampment of the whole. They compensate, however, for injuries done to seeds by the amazing quantities of insects and noxious vermin which they devour, in preference to grain.

The Raven is an inhabitant not only of our island, but also of most other parts discovered of the New World. Amongst the ancients it was thought to be a bird of much importance in augury, and the various modulations of its voice were studied with the most earnest attention, and were too often used by designing men to delude the unwary. It frequents the neighbourhood of great towns, where it is useful by carrying off the carrion and filth which it can scent at a great distance. It is a cunning bird and generally careful in keeping beyond the reach of a gun. When brought up young he becomes very familiar, and in a domestic state, possesses many qualities, that render him highly amusing. Busy and inquisitive he goes every where, affronts and drives off the dogs, plays his tricks on the poultry, and cultivates the good will of the cook-maid, who, is generally his favourite in the family.

Q q

CORNISH CROW.

THE Corvus Graculus or Cornish Crow is the moſt singular bird of that kind residing in England, being only found in the county of Cornwall. It is considerably larger in size than the common Crow, and of a fine blue or purple black; its legs are described by Mr. Pennant as of a bright and deep orange colour. The bill is large and much hooked; however, its whole appearance is rather elegant; being tender in their nature they are not able to bear stormy weather, but frequently remove from their abode for a week at a time. They are much taken with glittering objects, and very apt to snatch up bits of lighted sticks, so that instances have occurred of their setting houses on fire. The injury done by this Crow to thatched cottages is very great; for tearing holes into them with its long bill, in search of worms and insects, the rain is admitted and quickens their decay: it will also often pick out lime from walls in search of spiders and flies. The Cornish peasantry attend a good deal to the taming of them, although their noise is more shrill than the Jackdaw. When any thing strange or frightful appears they shriek out aloud; but when applying for food, or desirous of pleasing those who usually fondle them, their chattering is very soft and engaging. When tame they are very docile and amusing, and extremely regular to their time of feeding; but, however familiar they may be to their immediate friends, they will not admit a stranger to touch them. Their nests are built about the middle of the cliffs; the eggs, which are four or five, somewhat longer than the Jackdaw's and of a cinereous white with dusky blotches. From their being very tender they are seldom seen abroad but in very fine weather.

CINEREOUS CROW.

THE American or Cinereous Crow, a bird most commonly known in Canada or Hudson's Bay, is smaller than most of the rest of its compeers only weighing about four ounces. Its plumage is a brown grey, the feathers are very long, and in general so much unwebbed, as in many parts of the body to resemble hair. It is very often troublesome to the English and Indian Hunters, sometimes following them for a whole day; it will perch on a tree while a hunter is baiting his Martin traps, and as soon as his back is turned, go and eat the baits. If tamed, which is not a very difficult task, from the moment it is caught (notwithstanding a seeming appetite and an imitative power of sound, which it frequently exerts) death is always the result of confinement. This bird, when alive and in its native state has a strong tendency to boldness and freedom of behaviour. When urged by a strong desire of food, he attends closely upon houses and tents, and is so much inclined to pilfering, that no provisions, salt or fresh, is secure from his depredations. He is sufficiently bold to come into the tents, sit on the edge of the kettle when hanging over the fire and steal victuals out of the dishes. This anxiety for food consists of laying up a stock of provisions against winter, when no fruits are to be had abroad, in consequence of snow and the severity of the season, and is a remarkable instance of foresight in the bird tribe; for this sort of propensity is generally very uncommon amongst them. Its nest is mostly built in trees in the manner of the Blackbird and Thrush, and the female lays four blue eggs, but seldom hatches more than three, building very early in the spring, that it may have a longer and pleasanter period in the summer.

The JAY, *or painted* CROW.

THE beautifully tufted Jay is well known in the woods of England; it builds in the trees an artless nest, composed of sticks, fibres, and twigs, in which it lays five or six eggs. Its colours are placed in such a pleasing and various profusion as hardly to be conceived. Blue wing, coverts barred with black and white, with a delicate cinnamon-coloured back and breast, render it the most elegant of the Crow kind, to which we may add a beautiful tuft of white feathers on its forehead. When kept in a domestic state, the Jay may be made very familiar, and will catch and repeat a variety of sounds. One of them has been heard·to imitate so exactly the noise made by the action of the saw, as to induce passengers to suppose that a carpenter was at work in the house. A Jay kept by a person in the North of England had learned, at the approach of cattle, to set a Cur dog upon them, by whistling and calling him by his name. One day in winter, during a severe frost, the dog was by this means, excited to attack a Cow, that was big with calf, when the animal fell upon the ice and was much hurt. The Jay was complained of as a nuisance, and the owner was obliged to destroy it. The young birds continue in company of the parents until the next season, a strong instance of mutual affection. They feed on acorn nuts, seeds, and fruit of all kinds. In the vicinity of their resort, they frequently are found rather injurious to gardens in devouring gooseberries, cherries, and peas; but surely their great beauty might fairly entitle them to a small share of those bounties of nature, which are withheld and spoiled and thrown away, by the exorbitant highness of their price.

The ROOK.

THE Rook is about the size of a Carrion Crow, but its plumage is very glossy ; it also differs in having the nostrils and root of the bill naked, which in the Crow are covered with hairs; and this is supposed to arise from the Rook's continually thrusting its bill into the ground in search of worms and beetles. Besides these insects the Rook will frequently feed upon the Cockchafer in the grub state, and by that circumstance clear the fields of farmers, from a most destructive and secret enemy, which can hardly be exterminated in any other way. These birds are gregarious, and seen sometimes in flocks so numerous, that their cawing noise, gives to some visitors of the country a pleasant rural sound. Mr. White, in his History of Selburne, speaking of the evening exercises of Rooks, remarks, that just before dusk, they return in long strings from the foraging of the day, and rendezvous by thousands over Selburne Down, where they wheel round and dive in a playful manner in the air, exerting their voices; which being softened by the distance, become a pleasing murmur, not unlike the cry of a pack of hounds in deep echoing woods. When this ceremony is over, they retire to the deep woods of Tisted and Kepley. We remember, says Mr. White, a little girl who used to remark on such an occurrence in the true spirit of physico-theology, that the Rooks were saying their prayers, and yet this child was much too young to be aware, that the Scriptures have said, that " He feedeth the Ravens when they call upon him." In the wilds of Hampshire adjoining to the New Forest, after the Rook has reared his progeny, and has carried off such of them as have escaped the arts of men and boys, he retires every evening at a late hour during the autumn and winter months, to the closer coverts of the forest, having spent the day in the open fields and inclosures in quest of food. His late retreat to the forest is characteristic of the near approach of night.

> " Retiring from the downs, where all day long
> " They pick their scanty fare, a black'ning train
> " Of loit'ring Rooks, thick urge their weary flight,
> " And seek repose and shelter from the grave."

But although the forest may be called his winter habitation, he, generally every day, visits his nursery, preserving the idea of a family ; which, early in the spring, he begins to make provision for. Among all the sounds of animal nature, few are more pleasing than the cawing of Rooks. The Rook has but two or three notes, and when he attempts a *solo* we cannot praise his song; but when he sings in concert, which is his chief delight, these notes, although rough of themselves, being intermixed with those of the multitude, have, as it were, all their rough edges worn off, and become harmonious, especially when softened in the air, where the bird chiefly performs. We have this music in perfection when the whole colony is roused by the discharge of a gun, resembling, in a great degree, Handel's chorus of Te Deum. A remarkable circumstance, respecting these birds, occurred a few years ago, at Dallam, in Westmoreland, the seat of Daniel Wilson, Esq. there were two groves adjoining the park; one of which had, for many years, been the resort of a number of Herons, that, regularly every year, built and bred there. In the other was a very large Rookery ; for a long time the two tribes lived peaceably together; at length, in the spring of 1775, the trees of the Heronry were cut down and the young brood perished by the fall of the timber. The parent birds, not willing to be driven from the place, endeavoured to effect a settlement in the Rookery. The Rooks made an obstinate resistance ; but, after a desperate conflict, in the course of which many of the Rooks and some of the Herons lost their lives, the latter succeeded in obtaining possession of some of the trees, and that very spring built their nests afresh. The next season a similar conflict took place, which, with the former, was terminated by the victory of the Herons. Since that time victory has been superseded by a peace, which seems to have been agreed upon between them; the Rooks have relinquished part of the grove to the Herons, to which part alone they confine themselves, and the two communities appear to live together in as much harmony as before the dispute. The following anecdote of this sagacious community is related by Dr. Percival, in his Dissertation : A large colony of Rooks had subsisted many years in a grove, on the banks of the Irwell, near Manchester ; one serene evening, I placed myself within the view of it, and marked, with attention, the various labours, pastimes, and evolutions of this crowded society. The idle members amused themselves with chasing each other through endless mazes, and in their flight they made the air resound with an infinitude of discordant notes. In the midst of these playful exertions it unfortunately happened, that one Rook, by a sudden turn, struck his beak against the other, cutting part of his wing. The sufferer instantly fell into the river ; a general cry of distress ensued. The birds hovered, with every expression of anxiety, over their deceased companion. Animated by their sympathy, and perhaps by the language of counsel known to themselves, he sprang into the air by one strong effort, and reached the point of a rock, which projected into the water. The joy became loud and universal; but, alas! it was soon changed into notes of lamentation, for the unfortunate bird in attempting to fly towards his nest, dropped into the river and was drowned amidst the moans of the whole fraternity. There seems to be a wonderful antipathy between these birds and the Raven. Mr. Marthwick observes, that a Raven built its nest in a tree adjoining to a very numerous Rookery, all the Rooks immediately forsook the spot and have not returned to build there since. This is but a common occurrence, since the Raven will not suffer any bird to come within a quarter of a mile of its nest, being very fierce in defending it. It besides seizes the young Rooks from their nests to feed its own young, and the terror arising in the parent Rook from this consideration, induces them immediately to seek a safer situation.

[This page was blank in the original]

SPOTTED HYENA.

Drawn & Engrav'd by T.L. Busby, from Pollard Menagore.

Pub.d by J. Stratford, Sep.t 1. 1811.

PLATE 81

Genus—HYÆNA. THE SPOTTED HYÆNA.

Character.—Dog, with a head sloping towards the ground; the nose, body, and legs striped, in the *common* species with *brown* stripes, but in the present instance with *brown* and *black spots*, irregularly placed over the body; and of different distances from each other. The general length is four or five feet; tail short in comparison with other animals, and bushy only at the extremity. Feet and teeth like the Wolf; a large lump or hunch, something like that of a Bison, and which rests upon the neck and body jointly. Disposition highly fierce and untameable.

IN the wild and desolate regions of Africa, where the Lion and the Tiger, frequent the Desarts, the common or Striped Hyæna abounds: in every part of that extensive quarter of the globe it extends its ravages amongst the cattle. These ravenous and powerful depredators being of a cowardly, though fierce nature, seldom appear during the day, but in the midnight, when all is hushed in solemn silence, suddenly a dreadful howl is heard, echoing through the forest, the shepherd seizes his spear or gun and joins his comrades, by a loud whistle. Judging by the strength of their howling, and by their glittering and terrific eyes, he conceives their situation, and generally fires in succession. Bonfires are also immediately lighted by the women, who pursue and throw amongst them the blazing torches; thus being terrified by the appearance of flame, they generally disperse. As the morning, however, advances, they are pursued into the narrow haunts of the mountains, and the Hottentots, from their ex-

R r

[541]

cellent aim with the gun, and which the Europeans have for a long time introduced, have happily thinned the country.

Africa possesses only two species of the Hyæna Class, distinguished by the striped and the spotted character, and a difference of size. The latter is much the rarest of the two, and six inches shorter in the body than the first. It was fortunately brought from Africa alive, and from which specimen the drawing now offered was taken by Mr. Busby. Mr. Polito, of Exeter 'Change, has, in this instance, given a fresh proof of his munificence, in allowing this object to be copied; which shews a rising taste for natural knowledge. These fierce animals seem intended by Providence, by their rapacious qualities, to lessen the amazing number of Goats, Sheep, Antelopes, &c. which have now only small patches of fertile ground, in the burning plains and desarts of that great and varying region. Man's invention has not been allowed hitherto to destroy all the destructive animals; but Providence has wisely placed upon them a slowness and smallness of numbers in their births, and in the natural hatred of man, which considerably diminishes their increase.

Drawn by G. Perry. Engrav'd by T.L.Busby.

PAPILIO VOLCANICA.

Pub.d by J. Stratford, Sep.t 1.st 1811.

PLATE 82

[544]

Genus—PAPILIO. *Species*—PAPILIO VOL- CANICA. *Division*—ORBATUS.

Character. Form of the wings all rounded in their shape, and strongly elevated in the shoulders of the upper wings; the edges of the wings slightly undulated. The under wings of the fly, of a round form like the upper wings; distinctly separated from the upper wings, and have an in- dented border. The shades and colours of the fly under- neath, differing from the upper. The skin bright yellow in most of this division.

THE division of Orbatus forms one of the most beauti- ful of the *natural forms,* chosen by the Author as a strong in- dicative designation, a principal reason for differing from Fa- bricius, Linnæus, and other writers upon this subject, so am- ply described in the ARCANA. The Papilio Volcanica is a native of the Rio de la Plata in South America, and of Peru. The back is yellow, spotted with black and brown spots, irre- gularly placed. The underside coloured with the most splen- did orange tints, mixed with red, and of a light brown; small and large circular pellets, of a pale brown, and scattered une- qually, with black netted lines which cover the skin. The whole texture of this curious fly exhibits to a fanciful and ca- pricious mind the representation of a Volcano or Burning Mountain, with balls of fire, and burning streams of lava, rising amongst the smoke. The lower fly resembles a soil of a sulphur tint of a brown and burnt appearance, with several orifices placed at intervals. If imagination can be indulged so far, we may call to mind the description of Sir William Ha- milton, in his travels in the vicinity of Mount Vesuvius, at the time of an eruption. Balls of fire were seen spreading over the illuminated scene, mounting high in the air, and falling near to his feet. Rivers of red hot lava, running in red lines, were stopped for some time by a valley, then taking its course through lakes and pools, till it arrived at last at the sea, and buried its heat in the deep.

The Papilio Volcanica is a rare Butterfly, and the only one perhaps that has been seen in London, or in any private collection. The Entomological specimens of the British Museum are so withered and decayed, by the ruthless hand of Time and exposure to the Sun's beams, that but little information on that subject can be drawn from that source. The feathered parts of the wings of Butterflies and Moths seem by Nature fitted for decay, but the Engraver and Painter has power to transmit their image, and preserve the idea of all their colours for ages ; but Nature, in the dead flies, by contracting the membranes of the wings, breaks off and displaces the feathers on the head, body, and all other parts of the interesting and beautiful objects. A close drawer, covered with glass in one solid piece, with a small quantity of camphor, is found to be the best security against small insects. Add to this, the cabinet should be placed in a room into which the sun never shines ; being substituted by a walled sky-light. In the time of winter, when the weather is very damp, they should now and then be aired near a fire ; and by these simple precautions they may be well preserved for two or three hundred years.

[This page was blank in the original]

Drawn & Engrav'd by T.L.Busby from Polito's Menagerie.

CASSOWARY.

Pub.^d by J.Stratford, Sepr 1.st 1811.

PLATE 83

Natural Division.—GRALLÆ, or OSTRICH. CASSOWARA EXIMIA.

THE Neck bare of feathers, and a leather covering similar to the wattle of a Turkey, investing the neck and face, and of various colours, graduated in their shades from Orange to Red, Sea-green, or Purple. The feathers large, purple and dark brown, resembling, when plucked, and closely examined, three or four dark horse hairs, joined to a hollow quill which is very long, thin, and bushy, tufted and imbricated most extraordinarily, the whole leg very bare and scaly. The head crowned with a sort of cock's-comb of a grey colour; three toes on each foot, armed with claws, one external, another in the middle (both short). But what is curious is, the innermost being much longer than the two others. Bill long and sharply hooked.

This Cassowary or Cassowara has certainly a strong title to the character of beauty and singularity, and is reported to be found in South America, though it is rather uncertain, until more is known of its history, for at present it is very rare. The natives also are said to pronounce it *Cassowara*; and this is our authority for doing the same, and rejecting the *y*. We have given it to the class of birds of Emu or Ostrich; to the Turkey and the English Bustard it is distantly similar. As from its general appearance it seems not to be yet fully grown, we cannot positively ascertain whether it is larger or less than the East Indian kind. We shall give to our Readers a description of the English Bustard, by which he will be more able to form a comparison with the present bird. The Otistarda, or Great English Bustard, is the largest land fowl produced in our island, the male often weighing twenty-five pounds and upwards. The length is near four feet, and the head and neck are ash-coloured : the tail is shaded with bars of red and black. The female is much less in height than the male ; the top of the head is of a deep orange, and the rest of the head brown. Her colours are not so bright as those of the male ; and she is without the tuft on each side of the head. This bird is furnished with a sack or pouch, situated, with a support, in the fore part of the neck, and capable of containing two quarts of water; the entrance to which is immediately under the tongue. Thus living in extensive plains and commons, he can supply himself like the Ostrich upon all necessary occasions. This singular reservoir was first discovered by Dr. Douglas, who supposes that it not only uses it on occasions of thirst, but that he likewise uses it in defending himself from birds of prey upon any sudden attack, and then throws out the water with violence, as not unfrequently to baffle the pursuit of its enemy. This bird makes no nest ; they are found in Dorsetshire, as far as the Wolds in Yorkshire, and are often seen on Salisbury Plain. They are slow in taking wing, but run with great rapidity. The hen generally lays her eggs in a hole in the ground : these are two in number, as big as those of a Goose, and of a pale olive brown, marked with spots of a darker colour. If during her absence from her nest, any one should handle or even breath upon the eggs, she immediately abandons them. The young follow the dam soon after they are excluded from the egg, but are not capable for some time of flying. Bustards, in general, feed upon green corn, the tops of turnips, and various other vegetables, as well as on worms ; but they have been known also to eat frogs, mice, and young birds of the smaller kind, which they swallow whole. They are remarkable for their great timidity : they are now and then met with in flocks of fifty or more. They are not unfrequently hunted with small dogs or greyhounds.

[This page was blank in the original]

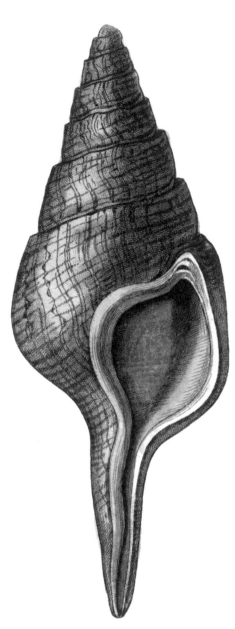

Drawn by G.Perry   Engrav'd by T.L.Busby

PINUS.

Pub.d by J.Stratford. Sep.t 1.st 1811.

PLATE 84

[552]

CONCHOLOGY.

Class—FOSSILIA. Order—UNIVALVÆ. Species—ROSTELLARIA.

Character.—Shells perfectly formed, of a pale brown colour, found in rocks and mountains of the earth.

THE instances which we have alluded to, in a former Part of the Arcana, of curious fossil shells, found in the neighbourhood of Paris, enclosed in beds of a light sandy chalk, will always be considered as one of the most astonishing facts of Nature. We have therefore in the present Number renewed this interesting subject, by the drawing which is hitherto unnamed; but appears to belong to the Genus Rostellaria, which has been separated from the Strombus, by having a straighter and longer beak, and a different mouth and spire. Dr. Jamieson has informed his friends of several mineralogical circumstances not hitherto explored, in the Islands of New Holland and Van Diemen's Land; and in traversing the most distant regions of the Hawsbury River, he had discovered the substances of coal, limestone, and fossil shells. But whether these were exactly similar to those substances, as extant at present in our own country, he had no means of ascertaining, to all appearances the coal is much richer and blacker; the lime much the same, the forms of the shells are wholly unknown.

Fossil shells may be said to have always one distinguishing characteristic; having lost in the earth their original native colour, they appear of a dull brown, sometimes inclining to a reddish brown, and in some few which are in the possession of Mr. Sowerby, the bands or marks are very faintly distinguishable. In the slate mountains of Switzerland are found petrifactions of fish, inclosed in a sort of white, pale, stony mixture, which form an exact representation of what the fish have formerly been, and the fins not at all obliterated. The fractures in the ribs or vertebræ prove the suddenness of the flood which has borne these creatures away, and placed them in the highest mountains. The variety of Fossils, may be pretty generally under the following heads, and as a chief outline of their particular geography. In England, Oysters and Cockles of a perfectly different form to any known at present are discovered. Ireland produces the Fossil Elk, which is also such as any other country cannot shew at present. America is not without its wonders. The tusks of an amazing Boar are found in Carolina only, the Boar must have been twenty feet long at least, to have sustained a tusk four feet long, and nine feet thick. The American Fossil Elephant must have differed from those of Asia in the form and size of its *tusks* and teeth, the tusk being as thick to the very end, as it is at the root. This every one knows is not so in the *modern* Elephant. The Mammoth is found in North America in a Fossil state, and also obtained within Siberia and Northern Russia, which has led some Philosophers to suppose them to have been amphibious; but this is not reconcileable to its other characters. The Southern part of America yields the Fossil Megatherium, so well described by Dr. Shaw, and has a less size than the Northern Mammoth, though resembling it in other respects. Switzerland produces fish as well as Spain, Fossil fish from slate quarries, which are preserved in the finest manner in the British Museum. Hungary and Bohemia present us with the Cornu Ammonis, with Boars' teeth, and their bones in a Fossil state. Thus the globe presents us with a treasure of contemplation, and rewards us for the examination of terrestrial as well as celestial objects. So that we may with an eminent sacred writer, conclude in the following sentiments:

" O Lord! how wonderful are thy works!
" In Wisdom hast thou made them all.
" The Earth is full of thy goodness!
" A foolish man doth not understand them."

Remarks on the PENGUIN *of the South Seas, and other Marine Birds.*

THE Penguin tribe of birds, seem to fill up the space of an intergenus between the land and sea birds, their wings being too short for flying, their body and neck very long; they may be said to hold the same place in the Southern and Pacific Oceans, that the Puffin or Auk does in the Northern Seas; being only found in the frigid and temperate zones of the southern hemisphere. They resemble them in almost all their habits, walking erect, and being very stupid; also, in their mode of feeding, and of making their nests; they swim with great swiftness, and are fortified against cold weather by being very fat: they hatch their young in an upright position, and parade the shore in tribes, making a general cackle like Geese, but in a hoarser and deeper tone. Mr. Forster, who had the honour of attending Captain Cook round the world, has described their habits in a very curious and entertaining manner, and it being a bird seldom seen stuffed, and never alive in England, we shall communicate to our readers a part of his description, previously observing, that the largest we have ever seen in a stuffed state, was four feet two inches high, therefore being twice the size of Mr. Forster's, and the latter rested. We may safely suppose that there are at least two, and perhaps more species of this Genus, existing in those desolate regions. " The Crested Penguins are inhabitants of the South Sea Islands. They have the names of Hopping Penguins and Jumping Jacks, from their action of jumping out of the water, sometimes three or four feet, on meeting with any obstacle in their course; and indeed they frequently do this without any other apparent cause than the desire of

s s

advancing by that means. All the Penguins swim deep in the sea, sinking above the breast, the head and neck only appearing out of the water; and they row themselves along with their finny and short wings as with oars. This bird nevertheless, and although resembling a short bag in its form, is still beautiful in its general appearance, the bill is red, and the upper mandible curved at the end, the lower one obtuse. The head, neck, back, and sides, are black, the edges of the wings and inside white ; they are of an orange colour ; the claws dusky ; but the female is destitute of the crest." Penrose describes a sort of Penguin that comes to certain places of the Falkland Islands, in incredible numbers, and lays its eggs; these places he tells us had become by their long residence entirely freed from grass, and he has given to them the name of Towns. The nests were composed of mud, raised into hillocks, about a foot high, and placed close to each other. Here the fancied streets and houses presented us with a sight that conveyed a most dreary, and I may say a most aweful idea that existed of these islands, being wholly deserted by the human species. A general stillness prevailed in these towns, and whenever we took a walk amongst them in order to provide ourselves with their eggs, we were regarded indeed with sidelong glances, but we carried no terror along with us. The eggs are larger than those of a Goose and are laid in pairs. When we took them once and sometimes twice in a season, they were as often replaced by the birds ; but prudence would not permit us to plunder too far, lest a future supply in the next year's brood might be prevented. They are very tenacious of life. Mr. Forster, left a great number of them apparently lifeless, from the blows they had received, while he went in pursuit of others; but they afterwards got up, and marched off with the utmost gravity. When attacked on every side, and of themselves numerous, they will all at once borrow courage from despair and bite violently the legs or c'oa hs of those who assail them.

The PUFFIN.

THE Puffin or Auk is a bird for the most part an inhabitant of the Northern Ocean; the Puffins seem to hold the same place in the Northern Parts of the World that the Penguin does in the South. They are birds of passage and are a full period of three years in completing their growth; and this has been discovered, by the changes which take place in their bills, and not building their nest until the third year. They are mostly seen to abound in Russia and Siberia, from which distant abode, they emigrate in the middle of April, to Scotland, Wales, Ireland, and the Isle of Man; their re-emigration taking place about the middle of August. At this period none but the youngest which are unfledged remain behind, and these become the prey of the Peregrine Falcon, as they come out of their holes, in the sandy hillocks of the sea shore. The business of their life, which seems their first object, is to form burrows in the crevices of rocks or soft earth and this they easily accomplish by their hooked bill, which very much resembles that of a large Parroquette. The work is chiefly the task of the males, who are so intent upon the business, as to suffer themselves at that time to be taken by the hand. Some, where there is a good opportunity, save themselves the trouble of forming holes by dispossessing the Rabbits of theirs. The food of the Puffins is fish or sea weed, which gives a taste of the herring; but those who have reconciled themselves to the peculiar flavour; and it being a bird instead of fish does not prevent in Lent and Easter the Catholics from presenting it at their tables, and when pickled with vinegar forming a most delicious dish. The Bill is black, flat, and compressed; the rest of the neck, grey and black; the legs of an orange colour, and they walk upright. In all these circumstances they strongly resemble the Penguin as before-mentioned.

Their eggs consist of a single one in each nest, and the males as well as females perform the office of incubation and warmth, relieving each other when they go to feed, and the first young ones are hatched in July. If the old ones should be attacked on their nest, it is said that they will first bite themselves, to shew their strong and vehement affection, and when released, instead of flying away, or deserting their young, they return to the burrows, with the fondest attention and regard.

The GANNET *or* SOLAN GOOSE.

THE Gannet or Solan Goose, is somewhat more than three feet in length, and weighs about seven pounds; the bill is six inches long; straight almost to the point, where it is a little bent; its edges are irregularly jagged, for the better se-curing of its prey; and about an inch from the upper mandible is a sharp process, pointing forward. The general colour of the plumage is a dirty white, with a cinereous tinge. Sur-rounding each eye there is a naked skin of fine blue. From the corner of the mouth a narrow slip of naked black skin extends to the hind part of the head, and beneath the chin is a pouch, capable of containing five or six Herrings, the neck is long, and the body flat and very full of feathers. On the crown of the head and the back part of the neck, is a small

buff colour space. The quill feathers and some other parts of
the wings are also black, except a green stripe. The tail is
wedge shaped, and consists of twelve sharp-pointed feathers.
These birds frequent several of the Hebrides, and are sometimes seen on the Cornish coast, but seldom occur in any
other parts of Europe. They are migratory, and first appear
in the above islands, about the month of March ; they remain
until August or September. These Geese are insatiable and
voracious, yet rather particular in respect to their prey ;
not deigning to eat any thing less than Mackarel or Herrings.
No fewer than one hundred thousand, are supposed to frequent the rocks of St. Kilda ; of which, including the young,
at least twenty thousand, are annually killed by the inhabitants for food. Allow that these birds remain in this part
about six months in the year, and that each bird destroys five
Herrings in a day ; which is considerably less than the average,
we have at least ninety millions of these, the finest fish in the
world, devoured annually by a single species—Saint Kilda
birds. They build their nests in the highest and steepest
rocks they can find near the sea, laying if undisturbed, only
one egg in the year ; but if that be taken away, they lay
another ; if that is *also* taken, a *third*, but never more in the
same year. The nests are composed of grass, sea plants,
or any refuse that they find floating in the water and fitted
for the purpose. The young during the first year differ
greatly from the old ones ; being of a dusky hue and speckled
with numerous triangular white spots. We are informed,
that when they pass from place to place, they unite in small
flocks of from five to fifteen, and, except in fine weather, fly
low, near the shore, but never fly over it, doubling the capes
and projecting parts, and keeping equally from the land.
During their fishing they rise high in the air, and soar aloft
and sail along, over the Herrings or Pilchards below, much
in the manner of Kites. When they observe the shoal crowded
thick together, they close their wings to their sides and pre-

cipitate themselves head foremost into the water, dropping almost like a stone. Their eye in this act is so correct, that they never fail to rise with a fish in their mouth.

Mr. Pennant observes that the natives of St. Kilda, one of the Hebrides, hold this bird in great estimation, and often undergo the greatest risks to obtain it. Where this is possible they climb up the rocks which they frequent, and in doing this they pass along paths so narrow and difficult, as to allow them barely room to cling, and that at an amazing height over a raging sea. Where this cannot be done, the fowler is lowered by a rope from the top, and to take the young often stations himself on the most dangerous ledges. Unterrified, however, he ransacks all the nests within his reach; and then by means of a pole and a rope moves to other places to do the same *dangerous* office.

We are told also, that where the old birds are to be taken, the inhabitants fasten a Herring to a bird-limed board and set it afloat, so that by falling furiously upon it, the bird may break its neck in the attempt; the Gannet seems to attend upon the Herrings and Pilchards all round the British Islands, and sometimes migrates in quest of food as far as Lisbon and the Tagus River, in the month of September. From this time till March it is not known what *becomes* of them. The young birds and the eggs are very pleasant eating; but the old ones are tough and unseemly. They sometimes have driven their bill through a place where the people of Cornwall were curing Pilchards, an inch and quarter thick, and by this means killed themselves upon the spot.

The Fishing CORMORANT of China.

THIS curious bird, is trained by the ingenious adaptations of the Chinese, to dive into lakes and pools, in that curious and interesting country and bring up large fish. They seem to be as regularly tamed and educated, as Spaniels or Hawks are for other purposes, and one man can easily manage a large number, by having several at once placed in different boats. The fisherman carries them out into the lake, perched without any confinement and quite tame, to the commands of his master. When arrived at the proper place, he marks the first signal which is given, and each flies away in different directions, to assign the fullest scope of action. They hunt about, they plunge through the reeds, and rise again, untill they have at last found their prey. They then seize it with their bill by the middle and carry it to their own governor. But when fish chances to be too large, they give each other mutual assistance, one seizes it by the head, the other by the tail, and in this manner they carry it to the boat together ; there the boatman stretches out one of his long oars on which they perch, and being delivered of their burden, they again fly off to pursue their sport. When they are weary they are allowed to rest awhile, but they are never fed till their work is over, as they would otherwise at once satiate themselves and discontinue the pursuit, the moment they had filled their stomachs with the food.

[This page was blank in the original]

READER'S GUIDE TO
PERRY'S *ARCANA*

INDEX

George Perry Jr. (1771-?) was an English architect of the Georgian period. In addition to his periodical, the *Arcana*, he published a large book on shells, the *Conchology*, for which he is somewhat better known.

Richard Eugene Petit (1931-) is a leading scholar of malacology (the study of mollusks) and of its history. For over 45 years he has authored major scholarly works on taxonomy and bibliography as well as original scientific research. He has been a research associate at the Smithsonian Institution and the Field Museum, and is a former president of the American Malacological Society.

Paul Callomon scanned and digitally restored the text and plates of the facsimile from the original.

Kate Nichols designed and typeset the book in Adobe Garamond.

Charles Ault managed the production of the book and supervised its printing on woodfree, uncoated 115gsm Yulong Pure Cream stock by Everbest Printing Co. of Nan Sha, China.